Radmila Bulajich Manfrino
José Antonio Gómez Ortega
Rogelio Valdez Delgado

Inequalities

A Mathematical Olympiad Approach

Birkhäuser
Basel · Boston · Berlin

Autors:

Radmila Bulajich Manfrino
Rogelio Valdez Delgado
Facultad de Ciencias
Universidad Autónoma Estado de Morelos
Av. Universidad 1001
Col. Chamilpa
62209 Cuernavaca, Morelos
México
e-mail: bulajich@uaem.mx
valdez@uaem.mx

José Antonio Gómez Ortega
Departamento de Matemàticas
Facultad de Ciencias, UNAM
Universidad Nacional Autónoma de México
Ciudad Universitaria
04510 México, D.F.
México
e-mail: jago@fciencias.unam.mx

2000 Mathematical Subject Classification 00A07; 26Dxx, 51M16

Library of Congress Control Number: 2009929571

Bibliografische Information der Deutschen Bibliothek
Die Deutsche Bibliothek verzeichnet diese Publikation in der Deutschen Nationalbibliografie; detaillierte bibliografische Daten sind im Internet über <http://dnb.ddb.de> abrufbar.

ISBN 978-3-0346-0049-1 Birkhäuser Verlag, Basel – Boston – Berlin

Basel · Boston · Berlin
Postfach 133, CH-4010 Basel, Schweiz
Ein Unternehmen von Springer Science+Business Media
Gedruckt auf säurefreiem Papier, hergestellt aus chlorfrei gebleichtem Zellstoff. TCF ∞

ISBN 978-3-0346-0049-1 e-ISBN 978-3-0346-0050-7

9 8 7 6 5 4 3 2 1 www.birkhauser.ch

BIRKHÄUSER

Introduction

This book is intended for the Mathematical Olympiad students who wish to prepare for the study of inequalities, a topic now of frequent use at various levels of mathematical competitions. In this volume we present both classic inequalities and the more useful inequalities for confronting and solving optimization problems. An important part of this book deals with geometric inequalities and this fact makes a big difference with respect to most of the books that deal with this topic in the mathematical olympiad.

The book has been organized in four chapters which have each of them a different character. Chapter 1 is dedicated to present basic inequalities. Most of them are numerical inequalities generally lacking any geometric meaning. However, where it is possible to provide a geometric interpretation, we include it as we go along. We emphasize the importance of some of these inequalities, such as the inequality between the arithmetic mean and the geometric mean, the Cauchy-Schwarz inequality, the rearrangement inequality, the Jensen inequality, the Muirhead theorem, among others. For all these, besides giving the proof, we present several examples that show how to use them in mathematical olympiad problems. We also emphasize how the substitution strategy is used to deduce several inequalities.

The main topic in Chapter 2 is the use of geometric inequalities. There we apply basic numerical inequalities, as described in Chapter 1, to geometric problems to provide examples of how they are used. We also work out inequalities which have a strong geometric content, starting with basic facts, such as the triangle inequality and the Euler inequality. We introduce examples where the symmetrical properties of the variables help to solve some problems. Among these, we pay special attention to the Ravi transformation and the correspondence between an inequality in terms of the side lengths of a triangle a, b, c and the inequalities that correspond to the terms s, r and R, the semiperimeter, the inradius and the circumradius of a triangle, respectively. We also include several classic geometric problems, indicating the methods used to solve them.

In Chapter 3 we present one hundred and twenty inequality problems that have appeared in recent events, covering all levels, from the national and up to the regional and international olympiad competitions.

In Chapter 4 we provide solutions to each of the two hundred and ten exercises in Chapters 1 and 2, and to the problems presented in Chapter 3. Most of the solutions to exercises or problems that have appeared in international mathematical competitions were taken from the official solutions provided at the time of the competitions. This is why we do not give individual credits for them.

Some of the exercises and problems concerning inequalities can be solved using different techniques, therefore you will find some exercises repeated in different sections. This indicates that the technique outlined in the corresponding section can be used as a tool for solving the particular exercise.

The material presented in this book has been accumulated over the last fifteen years mainly during work sessions with the students that won the national contest of the Mexican Mathematical Olympiad. These students were developing their skills and mathematical knowledge in preparation for the international competitions in which Mexico participates.

We would like to thank Rafael Martínez Enríquez, Leonardo Ignacio Martínez Sandoval, David Mireles Morales, Jesús Rodríguez Viorato and Pablo Soberón Bravo for their careful revision of the text and helpful comments for the improvement of the writing and the mathematical content.

Contents

Introduction **vii**

1 Numerical Inequalities **1**

1.1 Order in the real numbers 1
1.2 The quadratic function $ax^2 + 2bx + c$ 4
1.3 A fundamental inequality, arithmetic mean-geometric mean 7
1.4 A wonderful inequality: The rearrangement inequality 13
1.5 Convex functions 20
1.6 A helpful inequality 33
1.7 The substitution strategy 39
1.8 Muirhead's theorem 43

2 Geometric Inequalities **51**

2.1 Two basic inequalities 51
2.2 Inequalities between the sides of a triangle 54
2.3 The use of inequalities in the geometry of the triangle 59
2.4 Euler's inequality and some applications 66
2.5 Symmetric functions of a, b and c 70
2.6 Inequalities with areas and perimeters 75
2.7 Erdős-Mordell Theorem 80
2.8 Optimization problems 88

3 Recent Inequality Problems **101**

4 Solutions to Exercises and Problems **117**

4.1 Solutions to the exercises in Chapter 1 117
4.2 Solutions to the exercises in Chapter 2 140
4.3 Solutions to the problems in Chapter 3 162

Notation **205**

Bibliography 207

Index 209

Chapter 1

Numerical Inequalities

1.1 Order in the real numbers

A very important property of the real numbers is that they have an order. The order of the real numbers enables us to compare two numbers and to decide which one of them is greater or whether they are equal. Let us assume that the real numbers system contains a set P, which we will call the set of positive numbers, and we will express in symbols $x > 0$ if x belongs to P. We will also assume the following three properties.

Property 1.1.1. *Every real number x has one and only one of the following properties:*

(i) $x = 0$,

(ii) $x \in P$ *(that is, $x > 0$),*

(iii) $-x \in P$ *(that is, $-x > 0$).*

Property 1.1.2. *If $x,\ y \in P$, then $x+y \in P$ (in symbols $x > 0,\ y > 0 \Rightarrow x+y > 0$).*

Property 1.1.3. *If $x,\ y \in P$, then $xy \in P$ (in symbols $x > 0,\ y > 0 \Rightarrow xy > 0$).*

If we take the "real line" as the geometric representation of the real numbers, by this we mean a directed line where the number "0" has been located and serves to divide the real line into two parts, the positive numbers being on the side containing the number one "1". In general the number one is set on the right hand side of 0. The number 1 is positive, because if it were negative, since it has the property that $1 \cdot x = x$ for every x, we would have that any number $x \neq 0$ would satisfy $x \in P$ and $-x \in P$, which contradicts property 1.1.1.

Now we can define the relation a **is greater than** b if $a - b \in P$ (in symbols $a > b$). Similarly, a **is smaller than** b if $b - a \in P$ (in symbols $a < b$). Observe that

$a < b$ is equivalent to $b > a$. We can also define that a **is smaller than or equal to** b if $a < b$ or $a = b$ (using symbols $a \leq b$).

We will denote by $\mathbb{R}$ the set of real numbers and by $\mathbb{R}^+$ the set P of positive real numbers.

Example 1.1.4. (i) *If $a < b$ and c is any number, then $a + c < b + c$.*

(ii) *If $a < b$ and $c > 0$, then $ac < bc$.*

In fact, to prove (i) we see that $a + c < b + c \Leftrightarrow (b + c) - (a + c) > 0 \Leftrightarrow b - a > 0 \Leftrightarrow a < b$. To prove (ii), we proceed as follows: $a < b \Rightarrow b - a > 0$ and since $c > 0$, then $(b - a)c > 0$, therefore $bc - ac > 0$ and then $ac < bc$.

Exercise 1.1. Given two numbers a and b, exactly one of the following assertions is satisfied, $a = b$, $a > b$ or $a < b$.

Exercise 1.2. Prove the following assertions.

(i) $a < 0, b < 0 \Rightarrow ab > 0$.

(ii) $a < 0, b > 0 \Rightarrow ab < 0$.

(iii) $a < b, b < c \Rightarrow a < c$.

(iv) $a < b, c < d \Rightarrow a + c < b + d$.

(v) $a < b \Rightarrow -b < -a$.

(vi) $a > 0 \Rightarrow \dfrac{1}{a} > 0$.

(vii) $a < 0 \Rightarrow \dfrac{1}{a} < 0$.

(viii) $a > 0, b > 0 \Rightarrow \dfrac{a}{b} > 0$.

(ix) $0 < a < b, 0 < c < d \Rightarrow ac < bd$.

(x) $a > 1 \Rightarrow a^2 > a$.

(xi) $0 < a < 1 \Rightarrow a^2 < a$.

Exercise 1.3. (i) If $a > 0$, $b > 0$ and $a^2 < b^2$, then $a < b$.

(ii) If $b > 0$, we have that $\frac{a}{b} > 1$ if and only if $a > b$.

The absolute value of a real number x, which is denoted by $|x|$, is defined as

$$|x| = \begin{cases} x & \text{if } x \geq 0, \\ -x & \text{if} x < 0. \end{cases}$$

Geometrically, $|x|$ is the distance of the number x (on the real line) from the origin 0. Also, $|a - b|$ is the distance between the real numbers a and b on the real line.

Exercise 1.4. For any real numbers x, a and b, the following hold.

(i) $|x| \geq 0$, and is equal to zero only when $x = 0$.

(ii) $|-x| = |x|$.

(iii) $|x|^2 = x^2$.

(iv) $|ab| = |a|\,|b|$.

(v) $\left|\frac{a}{b}\right| = \frac{|a|}{|b|}$, with $b \neq 0$.

Proposition 1.1.5 (Triangle inequality). *The triangle inequality states that for any pair of real numbers a and b,*

$$|a + b| \leq |a| + |b|\,.$$

Moreover, the equality holds if and only if $ab \geq 0$.

Proof. Both sides of the inequality are positive; then using Exercise 1.3 it is sufficient to verify that $|a + b|^2 \leq (|a| + |b|)^2$:

$$\begin{aligned} |a+b|^2 &= (a+b)^2 = a^2 + 2ab + b^2 = |a|^2 + 2ab + |b|^2 \leq |a|^2 + 2\,|ab| + |b|^2 \\ &= |a|^2 + 2\,|a|\,|b| + |b|^2 = (|a| + |b|)^2\,. \end{aligned}$$

In the previous relations we observe only one inequality, which is obvious since $ab \leq |ab|$. Note that, when $ab \geq 0$, we can deduce that $ab = |ab| = |a|\,|b|$, and then the equality holds. □

The **general form of the triangle inequality** for real numbers $x_1, x_2, \ldots, x_n$, is

$$|x_1 + x_2 + \cdots + x_n| \leq |x_1| + |x_2| + \cdots + |x_n|.$$

The equality holds when all x_i's have the same sign. This can be proved in a similar way or by the use of induction. Another version of the last inequality, which is used very often, is the following:

$$|\pm x_1 \pm x_2 \pm \cdots \pm x_n| \leq |x_1| + |x_2| + \cdots + |x_n|.$$

Exercise 1.5. Let x, y, a, b be real numbers, prove that

(i) $|x| \leq b \Leftrightarrow -b \leq x \leq b$,

(ii) $||a| - |b|| \leq |a - b|$,

(iii) $x^2 + xy + y^2 \geq 0$,

(iv) $x > 0,\ y > 0 \Rightarrow x^2 - xy + y^2 > 0$.

Exercise 1.6. For real numbers a, b, c, prove that

$$|a| + |b| + |c| - |a + b| - |b + c| - |c + a| + |a + b + c| \geq 0.$$

Exercise 1.7. Let a, b be real numbers such that $0 \leq a \leq b \leq 1$. Prove that

(i) $0 \leq \dfrac{b-a}{1-ab} \leq 1$,

(ii) $0 \leq \dfrac{a}{1+b} + \dfrac{b}{1+a} \leq 1$,

(iii) $0 \leq ab^2 - ba^2 \leq \dfrac{1}{4}$.

Exercise 1.8. Prove that if n, m are positive integers, then $\frac{m}{n} < \sqrt{2}$ if and only if $\sqrt{2} < \frac{m+2n}{m+n}$.

Exercise 1.9. If $a \geq b$, $x \geq y$, then $ax + by \geq ay + bx$.

Exercise 1.10. If x, $y > 0$, then $\sqrt{\frac{x^2}{y}} + \sqrt{\frac{y^2}{x}} \geq \sqrt{x} + \sqrt{y}$.

Exercise 1.11. (Czech and Slovak Republics, 2004) Let a, b, c, d be real numbers with $a + d = b + c$, prove that

$$(a-b)(c-d) + (a-c)(b-d) + (d-a)(b-c) \geq 0.$$

Exercise 1.12. Let $f(a,b,c,d) = (a-b)^2 + (b-c)^2 + (c-d)^2 + (d-a)^2$. For $a < b < c < d$, prove that

$$f(a,c,b,d) > f(a,b,c,d) > f(a,b,d,c).$$

Exercise 1.13. (IMO, 1960) For which real values of x the following inequality holds:

$$\frac{4x^2}{(1-\sqrt{1+2x})^2} < 2x + 9?$$

Exercise 1.14. Prove that for any positive integer n, the fractional part of $\sqrt{4n^2+n}$ is smaller than $\frac{1}{4}$.

Exercise 1.15. (Short list IMO, 1996) Let a, b, c be positive real numbers such that $abc = 1$. Prove that

$$\frac{ab}{a^5+b^5+ab} + \frac{bc}{b^5+c^5+bc} + \frac{ca}{c^5+a^5+ca} \leq 1.$$

1.2 The quadratic function $ax^2 + 2bx + c$

One very useful inequality for the real numbers is $x^2 \geq 0$, which is valid for any real number x (it is sufficient to consider properties 1.1.1, 1.1.3 and Exercise 1.2 of the previous section). The use of this inequality leads to deducing many other inequalities. In particular, we can use it to find the maximum or minimum of a quadratic function $ax^2 + 2bx + c$. These quadratic functions appear frequently in optimization problems or in inequalities.

One common example consists in proving that if $a > 0$, the quadratic function $ax^2 + 2bx + c$ will have its minimum at $x = -\frac{b}{a}$ and the minimum value is $c - \frac{b^2}{a}$. In fact,

$$ax^2 + 2bx + c = a\left(x^2 + 2\frac{b}{a}x + \frac{b^2}{a^2}\right) + c - \frac{b^2}{a}$$
$$= a\left(x + \frac{b}{a}\right)^2 + c - \frac{b^2}{a}.$$

Since $\left(x + \frac{b}{a}\right)^2 \geq 0$ and the minimum value of this expression, zero, is attained when $x = -\frac{b}{a}$, we conclude that the minimum value of the quadratic function is $c - \frac{b^2}{a}$.

If $a < 0$, the quadratic function $ax^2+2bx+c$ will have a maximum at $x = -\frac{b}{a}$ and its value at this point is $c-\frac{b^2}{a}$. In fact, since $ax^2+2bx+c = a\left(x + \frac{b}{a}\right)^2 + c - \frac{b^2}{a}$ and since $a\left(x + \frac{b}{a}\right)^2 \leq 0$ (because $a < 0$), the greatest value of this last expression is zero, thus the quadratic function is always less than or equal to $c - \frac{b^2}{a}$ and assumes this value at the point $x = -\frac{b}{a}$.

Example 1.2.1. *If x, y are positive numbers with $x + y = 2a$, then the product xy is maximal when $x = y = a$.*

If $x + y = 2a$, then $y = 2a - x$. Hence, $xy = x(2a - x) = -x^2 + 2ax = -(x - a)^2 + a^2$ has a maximum value when $x = a$, and then $y = x = a$.

This can be interpreted geometrically as *"of all the rectangles with fixed perimeter, the one with the greatest area is the square"*. In fact, if x, y are the lengths of the sides of the rectangle, the perimeter is $2(x + y) = 4a$, and its area is xy, which is maximized when $x = y = a$.

Example 1.2.2. *If x, y are positive numbers with $xy = 1$, the sum $x+y$ is minimal when $x = y = 1$.*

If $xy = 1$, then $y = \frac{1}{x}$. It follows that $x + y = x + \frac{1}{x} = \left(\sqrt{x} - \frac{1}{\sqrt{x}}\right)^2 + 2$, and then $x + y$ is minimal when $\sqrt{x} - \frac{1}{\sqrt{x}} = 0$, that is, when $x = 1$. Therefore, $x = y = 1$.

This can also be interpreted geometrically in the following way, *"of all the rectangles with area 1, the square has the smallest perimeter"*. In fact, if x, y are the lengths of the sides of the rectangle, its area is $xy = 1$ and its perimeter is $2(x + y) = 2\left(x + \frac{1}{x}\right) = 2\left\{\left(\sqrt{x} - \frac{1}{\sqrt{x}}\right)^2 + 2\right\} \geq 4$. Moreover, the perimeter is 4 if and only if $\sqrt{x} - \frac{1}{\sqrt{x}} = 0$, that is, when $x = y = 1$.

Example 1.2.3. *For any positive number x, we have $x + \frac{1}{x} \geq 2$.*

Observe that $x + \frac{1}{x} = \left(\sqrt{x} - \frac{1}{\sqrt{x}}\right)^2 + 2 \geq 2$. Moreover, the equality holds if and only if $\sqrt{x} - \frac{1}{\sqrt{x}} = 0$, that is, when $x = 1$.

Example 1.2.4. *If $a, b > 0$, then $\frac{a}{b} + \frac{b}{a} \geq 2$, and the equality holds if and only if $a = b$.*

It is enough to consider the previous example with $x = \frac{a}{b}$.

Example 1.2.5. *Given $a, b, c > 0$, it is possible to construct a triangle with sides of length a, b, c if and only if $pa^2 + qb^2 > pqc^2$ for any p, q with $p + q = 1$.*

Remember that a, b and c are the lengths of the sides of a triangle if and only if $a + b > c$, $a + c > b$ and $b + c > a$.

Let

$$Q = pa^2 + qb^2 - pqc^2 = pa^2 + (1-p)b^2 - p(1-p)c^2 = c^2p^2 + (a^2 - b^2 - c^2)p + b^2,$$

therefore Q is a quadratic function[1] in p and

$$\begin{aligned} Q > 0 \quad &\Leftrightarrow \quad \Delta = \left[(a^2 - b^2 - c^2)^2 - 4b^2c^2\right] < 0 \\ &\Leftrightarrow \quad \left[a^2 - b^2 - c^2 - 2bc\right]\left[a^2 - b^2 - c^2 + 2bc\right] < 0 \\ &\Leftrightarrow \quad \left[a^2 - (b+c)^2\right]\left[a^2 - (b-c)^2\right] < 0 \\ &\Leftrightarrow \quad [a+b+c]\,[a-b-c]\,[a-b+c]\,[a+b-c] < 0 \\ &\Leftrightarrow \quad [b+c-a][c+a-b][a+b-c] > 0. \end{aligned}$$

Now, $[b+c-a][c+a-b][a+b-c] > 0$ if the three factors are positive or if one of them is positive and the other two are negative. However, the latter is impossible, because if $[b+c-a] < 0$ and $[c+a-b] < 0$, we would have, adding these two inequalities, that $c < 0$, which is false. Therefore the three factors are necessarily positive.

Exercise 1.16. Suppose the polynomial $ax^2 + bx + c$ satisfies the following: $a > 0$, $a + b + c \geq 0$, $a - b + c \geq 0$, $a - c \geq 0$ and $b^2 - 4ac \geq 0$. Prove that the roots are real and that they belong to the interval $-1 \leq x \leq 1$.

Exercise 1.17. If a, b, c are positive numbers, prove that it is not possible for the inequalities $a(1-b) > \frac{1}{4}$, $b(1-c) > \frac{1}{4}$, $c(1-a) > \frac{1}{4}$ to hold at the same time.

---

[1] A quadratic function $ax^2 + bx + c$ with $a > 0$ is positive when its discriminant $\Delta = b^2 - 4ac$ is negative, in fact, this follows from $ax^2 + bx + c = a(x + \frac{b}{2a})^2 + \frac{4ac - b^2}{4a}$. Remember that the roots are $\frac{-b \pm \sqrt{b^2 - 4ac}}{2a}$, and they are real when $\Delta \geq 0$, otherwise they are not real roots, and then $ax^2 + bx + c$ will have the same sign; this expression will be positive if $a > 0$.

1.3 A fundamental inequality, arithmetic mean-geometric mean

The first inequality that we consider, fundamental in optimization problems, is the inequality between the arithmetic mean and the geometric mean of two non-negative numbers a and b, which is expressed as

$$\frac{a+b}{2} \geq \sqrt{ab}, \qquad \text{(AM-GM)}.$$

Moreover, the equality holds if and only if $a = b$.

The numbers $\frac{a+b}{2}$ and $\sqrt{ab}$ are known as the **arithmetic mean** and the **geometric mean** of a and b, respectively. To prove the inequality we only need to observe that

$$\frac{a+b}{2} - \sqrt{ab} = \frac{a+b-2\sqrt{ab}}{2} = \frac{1}{2}\left(\sqrt{a}-\sqrt{b}\right)^2 \geq 0.$$

And the equality holds if and only if $\sqrt{a} = \sqrt{b}$, that is, when $a = b$.

Exercise 1.18. For $x \geq 0$, prove that $1 + x \geq 2\sqrt{x}$.

Exercise 1.19. For $x > 0$, prove that $x + \frac{1}{x} \geq 2$.

Exercise 1.20. For x, $y \in \mathbb{R}^+$, prove that $x^2 + y^2 \geq 2xy$.

Exercise 1.21. For x, $y \in \mathbb{R}^+$, prove that $2(x^2 + y^2) \geq (x + y)^2$.

Exercise 1.22. For x, $y \in \mathbb{R}^+$, prove that $\frac{1}{x} + \frac{1}{y} \geq \frac{4}{x+y}$.

Exercise 1.23. For a, b, $x \in \mathbb{R}^+$, prove that $ax + \frac{b}{x} \geq 2\sqrt{ab}$.

Exercise 1.24. If a, $b > 0$, then $\frac{a}{b} + \frac{b}{a} \geq 2$.

Exercise 1.25. If $0 < b \leq a$, then $\frac{1}{8}\frac{(a-b)^2}{a} \leq \frac{a+b}{2} - \sqrt{ab} \leq \frac{1}{8}\frac{(a-b)^2}{b}$.

Now, we will present a geometric and a visual proof of the following inequalities, for x, $y > 0$,

$$\frac{2}{\frac{1}{x}+\frac{1}{y}} \leq \sqrt{xy} \leq \frac{x+y}{2}. \tag{1.1}$$

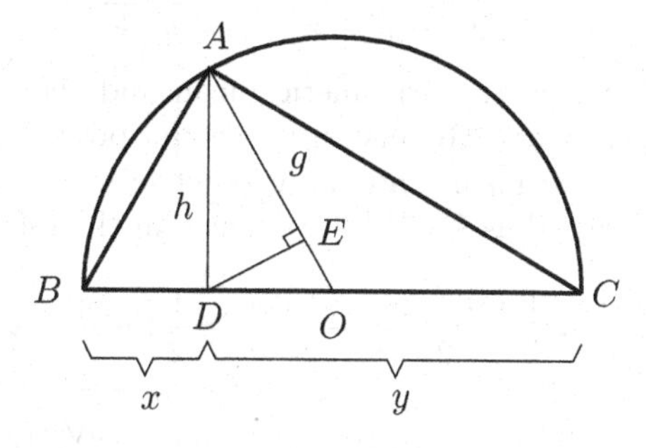

Let $x = BD$, $y = DC$ and let us construct a semicircle of diameter $BC = x + y$. Let A be the point where the perpendicular to BC in D intersects the semicircle and let E be the perpendicular projection from D to the radius AO. Let us write $AD = h$ and $AE = g$. Since ABD and CAD are similar right triangles, we deduce that

$$\frac{h}{y} = \frac{x}{h}, \qquad \text{then} \qquad h = \sqrt{xy}.$$

Also, since AOD and ADE are similar right triangles, we have

$$\frac{g}{\sqrt{xy}} = \frac{\sqrt{xy}}{\frac{x+y}{2}}, \qquad \text{then} \qquad g = \frac{2xy}{x+y} = \frac{2}{\left(\frac{1}{x} + \frac{1}{y}\right)}.$$

Finally, the geometry tells us that in a right triangle, the length of one leg is always smaller than the length of the hypotenuse. Hence, $g \leq h \leq \frac{x+y}{2}$, which can be written as

$$\frac{2}{\frac{1}{x} + \frac{1}{y}} \leq \sqrt{xy} \leq \frac{x+y}{2}.$$

The number $\frac{2}{\frac{1}{x}+\frac{1}{y}}$ is known as the **harmonic mean** of x and y, and the left inequality in (1.1) is known as the **inequality between the harmonic mean and the geometric mean.**

Some inequalities can be proved through the multiple application of a simple inequality and the use of a good idea to separate the problem into parts that are easier to deal with, a method which is often used to solve the following exercises.

Exercise 1.26. For $x, y, z \in \mathbb{R}^+$, $(x+y)(y+z)(z+x) \geq 8xyz$.

Exercise 1.27. For $x, y, z \in \mathbb{R}$, $x^2 + y^2 + z^2 \geq xy + yz + zx$.

Exercise 1.28. For $x, y, z \in \mathbb{R}^+$, $xy + yz + zx \geq x\sqrt{yz} + y\sqrt{zx} + z\sqrt{xy}$.

Exercise 1.29. For $x, y \in \mathbb{R}$, $x^2 + y^2 + 1 \geq xy + x + y$.

Exercise 1.30. For $x, y, z \in \mathbb{R}^+$, $\frac{1}{x} + \frac{1}{y} + \frac{1}{z} \geq \frac{1}{\sqrt{xy}} + \frac{1}{\sqrt{yz}} + \frac{1}{\sqrt{zx}}$.

Exercise 1.31. For $x, y, z \in \mathbb{R}^+$, $\frac{xy}{z} + \frac{yz}{x} + \frac{zx}{y} \geq x + y + z$.

Exercise 1.32. For $x, y, z \in \mathbb{R}$, $x^2 + y^2 + z^2 \geq x\sqrt{y^2 + z^2} + y\sqrt{x^2 + z^2}$.

The inequality between the arithmetic mean and the geometric mean can be extended to more numbers. For instance, we can prove the following inequality between the arithmetic mean and the geometric mean of four non-negative numbers a, b, c, d, expressed as $\frac{a+b+c+d}{4} \geq \sqrt[4]{abcd}$, in the following way:

$$\begin{aligned}\frac{a+b+c+d}{4} = \frac{1}{2}\left(\frac{a+b}{2} + \frac{c+d}{2}\right) &\geq \frac{1}{2}\left(\sqrt{ab} + \sqrt{cd}\right) \\ &\geq \sqrt{\sqrt{ab}\sqrt{cd}} = \sqrt[4]{abcd}.\end{aligned}$$

Observe that we have used the AM-GM inequality three times for two numbers in each case: with a and b, with c and d, and with $\sqrt{ab}$ and $\sqrt{cd}$. Moreover, the equality holds if and only if $a = b$, $c = d$ and $ab = cd$, that is, when the numbers satisfy $a = b = c = d$.

Exercise 1.33. For $x, y \in \mathbb{R}$, $x^4 + y^4 + 8 \geq 8xy$.

Exercise 1.34. For $a, b, c, d \in \mathbb{R}^+$, $(a + b + c + d)\left(\frac{1}{a} + \frac{1}{b} + \frac{1}{c} + \frac{1}{d}\right) \geq 16$.

Exercise 1.35. For $a, b, c, d \in \mathbb{R}^+$, $\frac{a}{b} + \frac{b}{c} + \frac{c}{d} + \frac{d}{a} \geq 4$.

A useful trick also exists for checking that the inequality $\frac{a+b+c}{3} \geq \sqrt[3]{abc}$ is true for any three non-negative numbers a, b and c. Consider the following four numbers a, b, c and $d = \sqrt[3]{abc}$. Since the AM-GM inequality holds for four numbers, we have $\frac{a+b+c+d}{4} \geq \sqrt[4]{abcd} = \sqrt[4]{d^3 d} = d$. Then $\frac{a+b+c}{4} \geq d - \frac{1}{4}d = \frac{3}{4}d$. Hence, $\frac{a+b+c}{3} \geq d = \sqrt[3]{abc}$.

These ideas can be used to justify the general version of the inequality for n non-negative numbers. If $a_1, a_2, \ldots, a_n$ are n non-negative numbers, we take the numbers A and G as

$$A = \frac{a_1 + a_2 + \cdots + a_n}{n} \quad \text{and} \quad G = \sqrt[n]{a_1 a_2 \cdots a_n}.$$

These numbers are known as the **arithmetic mean** and the **geometric mean** of the numbers $a_1, a_2, \ldots, a_n$, respectively.

Theorem 1.3.1 (The AM-GM inequality).

$$\frac{a_1 + a_2 + \cdots + a_n}{n} \geq \sqrt[n]{a_1 a_2 \cdots a_n}.$$

First proof (Cauchy). Let P_n be the statement $G \leq A$, for n numbers. We will proceed by mathematical induction on n, but this is an induction of the following type.

(1) We prove that the statement is true for 2 numbers, that is, P_2 is true.

(2) We prove that $P_n \Rightarrow P_{n-1}$.

(3) We prove that $P_n \Rightarrow P_{2n}$.

When (1), (2) and (3) are verified, all the assertions P_n with $n \geq 2$ are shown to be true. Now, we will prove these statements.

(1) This has already been done in the first part of the section.

(2) Let $a_1, \ldots, a_{n-1}$ be non-negative numbers and let $g = \sqrt[n-1]{a_1 \cdots a_{n-1}}$. Using this number and the numbers we already have, i.e., $a_1, \ldots, a_{n-1}$, we get n numbers to which we apply P_n,

$$\frac{a_1 + \cdots + a_{n-1} + g}{n} \geq \sqrt[n]{a_1 a_2 \cdots a_{n-1} g} = \sqrt[n]{g^{n-1} \cdot g} = g.$$

We deduce that $a_1+\cdots+a_{n-1}+g \geq ng$, and then it follows that $\frac{a_1+\cdots+a_{n-1}}{n-1} \geq g$, therefore P_{n-1} is true.

(3) Let $a_1, a_2, \ldots, a_{2n}$ be non-negative numbers, then

$$\begin{aligned} a_1+a_2+\cdots+a_{2n} &= (a_1+a_2)+(a_3+a_4)+\cdots+(a_{2n-1}+a_{2n}) \\ &\geq 2\left(\sqrt{a_1a_2}+\sqrt{a_3a_4}+\cdots+\sqrt{a_{2n-1}a_{2n}}\right) \\ &\geq 2n\left(\sqrt{a_1a_2}\sqrt{a_3a_4}\cdots\sqrt{a_{2n-1}a_{2n}}\right)^{\frac{1}{n}} \\ &= 2n\left(a_1a_2\cdots a_{2n}\right)^{\frac{1}{2n}}. \end{aligned}$$

We have applied the statement P_2 several times, and we have also applied the statement P_n to the numbers $\sqrt{a_1a_2}, \sqrt{a_3a_4}, \ldots, \sqrt{a_{2n-1}a_{2n}}$. □

Second proof. Let $A = \frac{a_1+\cdots+a_n}{n}$. We take two numbers a_i, one smaller than A and the other greater than A (if they exist), say $a_1 = A-h$ and $a_2 = A+k$, with $h, k > 0$.

We exchange a_1 and a_2 for two numbers that increase the product and fix the sum, defined as

$$a_1' = A, \quad a_2' = A+k-h.$$

Since $a_1'+a_2' = A+A+k-h = A-h+A+k = a_1+a_2$, clearly $a_1'+a_2'+a_3+\cdots+a_n = a_1+a_2+a_3+\cdots+a_n$, but $a_1'a_2' = A(A+k-h) = A^2+A(k-h)$ and $a_1a_2 = (A+k)(A-h) = A^2+A(k-h)-hk$, then $a_1'a_2' > a_1a_2$ and thus it follows that $a_1'a_2'a_3\cdots a_n > a_1a_2a_3\cdots a_n$.

If $A = a_1' = a_2' = a_3 = \cdots = a_n$, there is nothing left to prove (the equality holds), otherwise two elements will exist, one greater than A and the other one smaller than A and the argument is repeated. Since every time we perform this operation we create a number equal to A, this process can not be used more than n times. □

Example 1.3.2. *Find the maximum value of $x(1-x^3)$ for $0 \leq x \leq 1$.*

The idea of the proof is to exchange the product for another one in such a way that the sum of the elements involved in the new product is constant. If $y = x(1-x^3)$, it is clear that the right side of $3y^3 = 3x^3(1-x^3)(1-x^3)(1-x^3)$, expressed as the product of four numbers $3x^3$, $(1-x^3)$, $(1-x^3)$ and $(1-x^3)$, has a constant sum equal to 3. The AM-GM inequality for four numbers tells us that

$$3y^3 \leq \left(\frac{3x^3+3(1-x^3)}{4}\right)^4 = \left(\frac{3}{4}\right)^4.$$

Thus $y \leq \frac{3}{4\sqrt[3]{4}}$. Moreover, the maximum value is reached using $3x^3 = 1-x^3$, that is, if $x = \frac{1}{\sqrt[3]{4}}$.

Exercise 1.36. Let $x_i > 0$, $i = 1, \ldots, n$. Prove that

$$(x_1 + x_2 + \cdots + x_n)\left(\frac{1}{x_1} + \frac{1}{x_2} + \cdots + \frac{1}{x_n}\right) \geq n^2.$$

Exercise 1.37. If $\{a_1, \ldots, a_n\}$ is a permutation of $\{b_1, \ldots, b_n\} \subset \mathbb{R}^+$, then

$$\frac{a_1}{b_1} + \frac{a_2}{b_2} + \cdots + \frac{a_n}{b_n} \geq n \quad \text{and} \quad \frac{b_1}{a_1} + \frac{b_2}{a_2} + \cdots + \frac{b_n}{a_n} \geq n.$$

Exercise 1.38. If $a > 1$, then $a^n - 1 > n\left(a^{\frac{n+1}{2}} - a^{\frac{n-1}{2}}\right)$.

Exercise 1.39. If a, b, $c > 0$ and $(1+a)(1+b)(1+c) = 8$, then $abc \leq 1$.

Exercise 1.40. If a, b, $c > 0$, then $\frac{a^3}{b} + \frac{b^3}{c} + \frac{c^3}{a} \geq ab + bc + ca$.

Exercise 1.41. For non-negative real numbers a, b, c, prove that

$$a^2b^2 + b^2c^2 + c^2a^2 \geq abc(a + b + c).$$

Exercise 1.42. If a, b, $c > 0$, then

$$\left(a^2b + b^2c + c^2a\right)\left(ab^2 + bc^2 + ca^2\right) \geq 9a^2b^2c^2.$$

Exercise 1.43. If a, b, $c > 0$ satisfy that $abc = 1$, prove that

$$\frac{1 + ab}{1 + a} + \frac{1 + bc}{1 + b} + \frac{1 + ac}{1 + c} \geq 3.$$

Exercise 1.44. If a, b, $c > 0$, prove that

$$\frac{1}{a} + \frac{1}{b} + \frac{1}{c} \geq 2\left(\frac{1}{a+b} + \frac{1}{b+c} + \frac{1}{c+a}\right) \geq \frac{9}{a+b+c}.$$

Exercise 1.45. If $H_n = 1 + \frac{1}{2} + \cdots + \frac{1}{n}$, prove that

$$n(n+1)^{\frac{1}{n}} < n + H_n \quad \text{for} \quad n \geq 2.$$

Exercise 1.46. Let $x_1, x_2, \ldots, x_n > 0$ such that $\frac{1}{1+x_1} + \cdots + \frac{1}{1+x_n} = 1$. Prove that

$$x_1 x_2 \cdots x_n \geq (n-1)^n.$$

Exercise 1.47. (Short list IMO, 1998) Let $a_1, a_2, \ldots, a_n$ be positive numbers with $a_1 + a_2 + \cdots + a_n < 1$, prove that

$$\frac{a_1 a_2 \cdots a_n \left[1 - (a_1 + a_2 + \cdots + a_n)\right]}{(a_1 + a_2 + \cdots + a_n)(1 - a_1)(1 - a_2) \cdots (1 - a_n)} \leq \frac{1}{n^{n+1}}.$$

Exercise 1.48. Let $a_1, a_2, \ldots, a_n$ be positive numbers such that $\frac{1}{1+a_1}+\cdots+\frac{1}{1+a_n} = 1$. Prove that

$$\sqrt{a_1}+\cdots+\sqrt{a_n} \geq (n-1)\left(\frac{1}{\sqrt{a_1}}+\cdots+\frac{1}{\sqrt{a_n}}\right).$$

Exercise 1.49. (APMO, 1991) Let $a_1, a_2, \ldots, a_n, b_1, b_2, \ldots, b_n$ be positive numbers with $a_1+a_2+\cdots+a_n = b_1+b_2+\cdots+b_n$. Prove that

$$\frac{a_1^2}{a_1+b_1}+\cdots+\frac{a_n^2}{a_n+b_n} \geq \frac{1}{2}(a_1+\cdots+a_n).$$

Exercise 1.50. Let a, b, c be positive numbers, prove that

$$\frac{1}{a^3+b^3+abc}+\frac{1}{b^3+c^3+abc}+\frac{1}{c^3+a^3+abc} \leq \frac{1}{abc}.$$

Exercise 1.51. Let a, b, c be positive numbers with $a+b+c=1$, prove that

$$\left(\frac{1}{a}+1\right)\left(\frac{1}{b}+1\right)\left(\frac{1}{c}+1\right) \geq 64.$$

Exercise 1.52. Let a, b, c be positive numbers with $a+b+c=1$, prove that

$$\left(\frac{1}{a}-1\right)\left(\frac{1}{b}-1\right)\left(\frac{1}{c}-1\right) \geq 8.$$

Exercise 1.53. (Czech and Slovak Republics, 2005) Let a, b, c be positive numbers that satisfy $abc=1$, prove that

$$\frac{a}{(a+1)(b+1)}+\frac{b}{(b+1)(c+1)}+\frac{c}{(c+1)(a+1)} \geq \frac{3}{4}.$$

Exercise 1.54. Let a, b, c be positive numbers for which $\frac{1}{1+a}+\frac{1}{1+b}+\frac{1}{1+c}=1$. Prove that

$$abc \geq 8.$$

Exercise 1.55. Let a, b, c be positive numbers, prove that

$$\frac{2ab}{a+b}+\frac{2bc}{b+c}+\frac{2ca}{c+a} \leq a+b+c.$$

Exercise 1.56. Let $a_1, a_2, \ldots, a_n, b_1, b_2, \ldots, b_n$ be positive numbers, prove that

$$\sum_{i=1}^{n}\frac{1}{a_ib_i}\sum_{i=1}^{n}(a_i+b_i)^2 \geq 4n^2.$$

Exercise 1.57. (Russia, 1991) For all non-negative real numbers x, y, z, prove that

$$\frac{(x+y+z)^2}{3} \geq x\sqrt{yz}+y\sqrt{zx}+z\sqrt{xy}.$$

Exercise 1.58. (Russia, 1992) For all positive real numbers x, y, z, prove that

$$x^4 + y^4 + z^2 \geq \sqrt{8}xyz.$$

Exercise 1.59. (Russia, 1992) For any real numbers x, $y > 1$, prove that

$$\frac{x^2}{y-1} + \frac{y^2}{x-1} \geq 8.$$

1.4 A wonderful inequality: The rearrangement inequality

Consider two collections of real numbers in increasing order,

$$a_1 \leq a_2 \leq \cdots \leq a_n \quad \text{and} \quad b_1 \leq b_2 \leq \cdots \leq b_n.$$

For any permutation $(a'_1, a'_2, \ldots, a'_n)$ of $(a_1, a_2, \ldots, a_n)$, it happens that

$$a_1b_1 + a_2b_2 + \cdots + a_nb_n \geq a'_1b_1 + a'_2b_2 + \cdots + a'_nb_n \tag{1.2}$$

$$\geq a_nb_1 + a_{n-1}b_2 + \cdots + a_1b_n. \tag{1.3}$$

Moreover, the equality in (1.2) holds if and only if $(a'_1, a'_2, \ldots, a'_n) = (a_1, a_2, \ldots, a_n)$. And the equality in (1.3) holds if and only if $(a'_1, a'_2, \ldots, a'_n) = (a_n, a_{n-1}, \ldots, a_1)$. Inequality (1.2) is known as the **rearrangement inequality**.

Corollary 1.4.1. *For any permutation* $(a'_1, a'_2, \ldots, a'_n)$ *of* $(a_1, a_2, \ldots, a_n)$, *it follows that*

$$a_1^2 + a_2^2 + \cdots + a_n^2 \geq a_1a'_1 + a_2a'_2 + \cdots + a_na'_n.$$

Corollary 1.4.2. *For any permutation* $(a'_1, a'_2, \ldots, a'_n)$ *of* $(a_1, a_2, \ldots, a_n)$, *it follows that*

$$\frac{a'_1}{a_1} + \frac{a'_2}{a_2} + \cdots + \frac{a'_n}{a_n} \geq n.$$

Proof (of the rearrangement inequality). Suppose that $b_1 \leq b_2 \leq \cdots \leq b_n$. Let

$$S = a_1b_1 + a_2b_2 + \cdots + a_rb_r + \cdots + a_sb_s + \cdots + a_nb_n,$$

$$S' = a_1b_1 + a_2b_2 + \cdots + a_sb_r + \cdots + a_rb_s + \cdots + a_nb_n.$$

The difference between S and S' is that the coefficients of b_r and b_s, where $r < s$, are switched. Hence

$$S - S' = a_rb_r + a_sb_s - a_sb_r - a_rb_s = (b_s - b_r)(a_s - a_r).$$

Thus, we have that $S \geq S'$ if and only if $a_s \geq a_r$. Repeating this process we get the result that the sum S is maximal when $a_1 \leq a_2 \leq \cdots \leq a_n$. $\square$

Example 1.4.3. (IMO, 1975) *Consider two collections of numbers* $x_1 \leq x_2 \leq \cdots \leq x_n$ *and* $y_1 \leq y_2 \leq \cdots \leq y_n$, *and one permutation* $(z_1, z_2, \ldots, z_n)$ *of* $(y_1, y_2, \ldots, y_n)$. *Prove that*

$$(x_1 - y_1)^2 + \cdots + (x_n - y_n)^2 \leq (x_1 - z_1)^2 + \cdots + (x_n - z_n)^2.$$

By squaring and rearranging this last inequality, we find that it is equivalent to

$$\sum_{i=1}^{n} x_i^2 - 2\sum_{i=1}^{n} x_i y_i + \sum_{i=1}^{n} y_i^2 \leq \sum_{i=1}^{n} x_i^2 - 2\sum_{i=1}^{n} x_i z_i + \sum_{i=1}^{n} z_i^2,$$

but since $\sum_{i=1}^{n} y_i^2 = \sum_{i=1}^{n} z_i^2$, then the inequality we have to prove turns to be equivalent to

$$\sum_{i=1}^{n} x_i z_i \leq \sum_{i=1}^{n} x_i y_i,$$

which in turn is inequality (1.2).

Example 1.4.4. (IMO, 1978) *Let* $x_1, x_2, \ldots, x_n$ *be distinct positive integers, prove that*

$$\frac{x_1}{1^2} + \frac{x_2}{2^2} + \cdots + \frac{x_n}{n^2} \geq \frac{1}{1} + \frac{1}{2} + \cdots + \frac{1}{n}.$$

Let $(a_1, a_2, \ldots, a_n)$ be a permutation of $(x_1, x_2, \ldots, x_n)$ with $a_1 \leq a_2 \leq \cdots \leq a_n$ and let $(b_1, b_2, \ldots, b_n) = \left(\frac{1}{n^2}, \frac{1}{(n-1)^2}, \ldots, \frac{1}{1^2}\right)$; that is, $b_i = \frac{1}{(n+1-i)^2}$ for $i = 1, \ldots, n$.

Consider the permutation $(a'_1, a'_2, \ldots, a'_n)$ of $(a_1, a_2, \ldots, a_n)$ defined by $a'_i = x_{n+1-i}$, for $i = 1, \ldots, n$. Using inequality (1.3) we can argue that

$$\begin{aligned}
\frac{x_1}{1^2} + \frac{x_2}{2^2} + \cdots + \frac{x_n}{n^2} &= a'_1 b_1 + a'_2 b_2 + \cdots + a'_n b_n \\
&\geq a_n b_1 + a_{n-1} b_2 + \cdots + a_1 b_n \\
&= a_1 b_n + a_2 b_{n-1} + \cdots + a_n b_1 \\
&= \frac{a_1}{1^2} + \frac{a_2}{2^2} + \cdots + \frac{a_n}{n^2}.
\end{aligned}$$

Since $1 \leq a_1$, $2 \leq a_2$, $\ldots$, $n \leq a_n$, we have that

$$\frac{x_1}{1^2} + \frac{x_2}{2^2} + \cdots + \frac{x_n}{n^2} \geq \frac{a_1}{1^2} + \frac{a_2}{2^2} + \cdots + \frac{a_n}{n^2} \geq \frac{1}{1^2} + \frac{2}{2^2} + \cdots + \frac{n}{n^2} = \frac{1}{1} + \frac{1}{2} + \cdots + \frac{1}{n}.$$

Example 1.4.5. (IMO, 1964) *Suppose that* a, b, c *are the lengths of the sides of a triangle. Prove that*

$$a^2(b + c - a) + b^2(a + c - b) + c^2(a + b - c) \leq 3abc.$$

Since the expression is a symmetric function of a, b and c, we can assume, without loss of generality, that $c \le b \le a$. In this case, $a\,(b+c-a) \le b\,(a+c-b) \le c\,(a+b-c)$.

For instance, the first inequality is proved in the following way:

$$\begin{aligned} a\,(b+c-a) \le b\,(a+c-b) \quad &\Leftrightarrow \quad ab+ac-a^2 \le ab+bc-b^2 \\ &\Leftrightarrow \quad (a-b)\,c \le (a+b)\,(a-b) \\ &\Leftrightarrow \quad (a-b)\,(a+b-c) \ge 0. \end{aligned}$$

By (1.3) of the rearrangement inequality, we have

$$a^2(b+c-a)+b^2(c+a-b)+c^2(a+b-c) \le ba(b+c-a)+cb(c+a-b)+ac(a+b-c),$$

$$a^2(b+c-a)+b^2(c+a-b)+c^2(a+b-c) \le ca(b+c-a)+ab(c+a-b)+bc(a+b-c).$$

Therefore, $2\left[a^2(b+c-a)+b^2(c+a-b)+c^2(a+b-c)\right] \le 6abc$.

Example 1.4.6. (IMO, 1983) *Let a, b and c be the lengths of the sides of a triangle. Prove that*

$$a^2b(a-b)+b^2c(b-c)+c^2a\,(c-a) \ge 0.$$

Consider the case $c \le b \le a$ (the other cases are similar).

As in the previous example, we have that $a(b+c-a) \le b(a+c-b) \le c(a+b-c)$ and since $\frac{1}{a} \le \frac{1}{b} \le \frac{1}{c}$, using Inequality (1.2) leads us to

$$\begin{aligned} &\frac{1}{a}a(b+c-a)+\frac{1}{b}b(c+a-b)+\frac{1}{c}c(a+b-c) \\ &\qquad\qquad \ge \frac{1}{c}a(b+c-a)+\frac{1}{a}b(c+a-b)+\frac{1}{b}c(a+b-c). \end{aligned}$$

Therefore,

$$a+b+c \ge \frac{a\,(b-a)}{c}+\frac{b(c-b)}{a}+\frac{c\,(a-c)}{b}+a+b+c.$$

It follows that $\frac{a(b-a)}{c}+\frac{b(c-b)}{a}+\frac{c(a-c)}{b} \le 0$. Multiplying by abc, we obtain

$$a^2b\,(a-b)+b^2c(b-c)+c^2a(c-a) \ge 0.$$

Example 1.4.7 (Cauchy-Schwarz inequality). *For real numbers $x_1, \ldots, x_n, y_1, \ldots, y_n$, the following inequality holds:*

$$\left(\sum_{i=1}^{n} x_iy_i\right)^2 \le \left(\sum_{i=1}^{n} x_i^2\right)\left(\sum_{i=1}^{n} y_i^2\right).$$

The equality holds if and only if there exists some $\lambda \in \mathbb{R}$ with $x_i = \lambda y_i$ for all $i = 1, 2, \ldots, n$.

If $x_1 = x_2 = \cdots = x_n = 0$ or $y_1 = y_2 = \cdots = y_n = 0$, the result is evident. Otherwise, let $S = \sqrt{\sum_{i=1}^n x_i^2}$ and $T = \sqrt{\sum_{i=1}^n y_i^2}$, where it is clear that $S, T \neq 0$. Take $a_i = \frac{x_i}{S}$ and $a_{n+i} = \frac{y_i}{T}$ for $i = 1, 2, \ldots, n$. Using Corollary 1.4.1,

$$\begin{aligned} 2 &= \sum_{i=1}^{n} \frac{x_i^2}{S^2} + \sum_{i=1}^{n} \frac{y_i^2}{T^2} = \sum_{i=1}^{2n} a_i^2 \\ &\geq a_1 a_{n+1} + a_2 a_{n+2} + \cdots + a_n a_{2n} + a_{n+1} a_1 + \cdots + a_{2n} a_n \\ &= 2\frac{x_1 y_1 + x_2 y_2 + \cdots + x_n y_n}{ST}. \end{aligned}$$

The equality holds if and only if $a_i = a_{n+i}$ for $i = 1, 2, \ldots, n$, or equivalently, if and only if $x_i = \frac{S}{T} y_i$ for $i = 1, 2, \ldots, n$.

Another proof of the Cauchy-Schwarz inequality can be established using Lagrange's identity

$$\left(\sum_{i=1}^{n} x_i y_i\right)^2 = \sum_{i=1}^{n} x_i^2 \sum_{i=1}^{n} y_i^2 - \frac{1}{2}\sum_{i=1}^{n}\sum_{j=1}^{n}(x_i y_j - x_j y_i)^2.$$

The importance of the Cauchy-Schwarz inequality will be felt throughout the remaining part of this book, as we will use it as a tool to solve many exercises and problems proposed here.

Example 1.4.8 (Nesbitt's inequality). *For a, b, $c \in \mathbb{R}^+$, we have*

$$\frac{a}{b+c} + \frac{b}{c+a} + \frac{c}{a+b} \geq \frac{3}{2}.$$

Without loss of generality, we can assume that $a \leq b \leq c$, and then it follows that $a + b \leq c + a \leq b + c$ and $\frac{1}{b+c} \leq \frac{1}{c+a} \leq \frac{1}{a+b}$.

Using the rearrangement inequality (1.2) twice, we obtain

$$\begin{aligned} \frac{a}{b+c} + \frac{b}{c+a} + \frac{c}{a+b} &\geq \frac{b}{b+c} + \frac{c}{c+a} + \frac{a}{a+b}, \\ \frac{a}{b+c} + \frac{b}{c+a} + \frac{c}{a+b} &\geq \frac{c}{b+c} + \frac{a}{c+a} + \frac{b}{a+b}. \end{aligned}$$

Hence,

$$2\left(\frac{a}{b+c} + \frac{b}{c+a} + \frac{c}{a+b}\right) \geq \left(\frac{b+c}{b+c} + \frac{c+a}{c+a} + \frac{a+b}{a+b}\right) = 3.$$

Another way to prove the inequality is using Inequality (1.3) twice,

$$\begin{aligned} \frac{c+a}{b+c} + \frac{a+b}{c+a} + \frac{b+c}{a+b} &\geq 3, \\ \frac{a+b}{b+c} + \frac{b+c}{c+a} + \frac{c+a}{a+b} &\geq 3. \end{aligned}$$

Then, after adding the two expressions, we get $\frac{2a+b+c}{b+c} + \frac{2b+c+a}{c+a} + \frac{2c+a+b}{a+b} \geq 6$, therefore

$$\frac{2a}{b+c} + \frac{2b}{c+a} + \frac{2c}{a+b} \geq 3.$$

Example 1.4.9. (IMO, 1995) *Let a, b, c be positive real numbers with $abc = 1$. Prove that*

$$\frac{1}{a^3(b+c)} + \frac{1}{b^3(c+a)} + \frac{1}{c^3(a+b)} \geq \frac{3}{2}.$$

Without loss of generality, we can assume that $c \leq b \leq a$. Let $x = \frac{1}{a}$, $y = \frac{1}{b}$ and $z = \frac{1}{c}$, thus

$$\begin{aligned} S &= \frac{1}{a^3(b+c)} + \frac{1}{b^3(c+a)} + \frac{1}{c^3(a+b)} \\ &= \frac{x^3}{\frac{1}{y}+\frac{1}{z}} + \frac{y^3}{\frac{1}{z}+\frac{1}{x}} + \frac{z^3}{\frac{1}{x}+\frac{1}{y}} \\ &= \frac{x^2}{y+z} + \frac{y^2}{z+x} + \frac{z^2}{x+y}. \end{aligned}$$

Since $x \leq y \leq z$, we can deduce that $x+y \leq z+x \leq y+z$ and also that $\frac{x}{y+z} \leq \frac{y}{z+x} \leq \frac{z}{x+y}$. Using the rearrangement inequality (1.2), we show that

$$\begin{aligned} \frac{x^2}{y+z} + \frac{y^2}{z+x} + \frac{z^2}{x+y} &\geq \frac{xy}{y+z} + \frac{yz}{z+x} + \frac{zx}{x+y}, \\ \frac{x^2}{y+z} + \frac{y^2}{z+x} + \frac{z^2}{x+y} &\geq \frac{xz}{y+z} + \frac{yx}{z+x} + \frac{zy}{x+y}, \end{aligned}$$

which in turn leads to $2S \geq x+y+z \geq 3\sqrt[3]{xyz} = 3$. Therefore, $S \geq \frac{3}{2}$.

Example 1.4.10. (APMO, 1998) *Let a, b, $c \in \mathbb{R}^+$, prove that*

$$\left(1+\frac{a}{b}\right)\left(1+\frac{b}{c}\right)\left(1+\frac{c}{a}\right) \geq 2\left(1+\frac{a+b+c}{\sqrt[3]{abc}}\right).$$

Observe that

$$\begin{aligned} & \left(1+\frac{a}{b}\right)\left(1+\frac{b}{c}\right)\left(1+\frac{c}{a}\right) \geq 2\left(1+\frac{a+b+c}{\sqrt[3]{abc}}\right) \\ \Leftrightarrow\quad & 1+\left(\frac{a}{b}+\frac{b}{c}+\frac{c}{a}\right)+\left(\frac{a}{c}+\frac{c}{b}+\frac{b}{a}\right)+\frac{abc}{abc} \geq 2\left(1+\frac{a+b+c}{\sqrt[3]{abc}}\right) \\ \Leftrightarrow\quad & \frac{a}{b}+\frac{b}{c}+\frac{c}{a}+\frac{a}{c}+\frac{c}{b}+\frac{b}{a} \geq \frac{2(a+b+c)}{\sqrt[3]{abc}}. \end{aligned}$$

Now we set $a = x^3$, $b = y^3$, $c = z^3$. We need to prove that

$$\frac{x^3}{y^3} + \frac{y^3}{z^3} + \frac{z^3}{x^3} + \frac{x^3}{z^3} + \frac{z^3}{y^3} + \frac{y^3}{x^3} \geq \frac{2\left(x^3 + y^3 + z^3\right)}{xyz}.$$

But, if we consider

$$(a_1, a_2, a_3, a_4, a_5, a_6) = \left(\frac{x}{y}, \frac{y}{z}, \frac{z}{x}, \frac{x}{z}, \frac{z}{y}, \frac{y}{x}\right),$$
$$(a'_1, a'_2, a'_3, a'_4, a'_5, a'_6) = \left(\frac{y}{z}, \frac{z}{x}, \frac{x}{y}, \frac{z}{y}, \frac{y}{x}, \frac{x}{z}\right),$$
$$(b_1, b_2, b_3, b_4, b_5, b_6) = \left(\frac{x^2}{y^2}, \frac{y^2}{z^2}, \frac{z^2}{x^2}, \frac{x^2}{z^2}, \frac{z^2}{y^2}, \frac{y^2}{x^2}\right),$$

we are led to the following result:

$$\begin{aligned}\frac{x^3}{y^3} + \frac{y^3}{z^3} + \frac{z^3}{x^3} + \frac{x^3}{z^3} + \frac{z^3}{y^3} + \frac{y^3}{x^3} &\geq \frac{x^2}{y^2}\frac{y}{z} + \frac{y^2}{z^2}\frac{z}{x} + \frac{z^2}{x^2}\frac{x}{y} + \frac{x^2}{z^2}\frac{z}{y} + \frac{z^2}{y^2}\frac{y}{x} + \frac{y^2}{x^2}\frac{x}{z}\\ &= \frac{x^2}{yz} + \frac{y^2}{zx} + \frac{z^2}{xy} + \frac{x^2}{zy} + \frac{z^2}{yx} + \frac{y^2}{xz}\\ &= \frac{2\left(x^3 + y^3 + z^3\right)}{xyz}.\end{aligned}$$

Example 1.4.11 (Tchebyshev's inequality). *Let* $a_1 \leq a_2 \leq \cdots \leq a_n$ *and* $b_1 \leq b_2 \leq \cdots \leq b_n$, *then*

$$\frac{a_1b_1 + a_2b_2 + \cdots + a_nb_n}{n} \geq \frac{a_1 + a_2 + \cdots + a_n}{n} \cdot \frac{b_1 + b_2 + \cdots + b_n}{n}.$$

Applying the rearrangement inequality several times, we get

$$\begin{aligned}a_1b_1 + \cdots + a_nb_n &= a_1b_1 + a_2b_2 + \cdots + a_nb_n,\\ a_1b_1 + \cdots + a_nb_n &\geq a_1b_2 + a_2b_3 + \cdots + a_nb_1,\\ a_1b_1 + \cdots + a_nb_n &\geq a_1b_3 + a_2b_4 + \cdots + a_nb_2,\\ &\vdots\\ a_1b_1 + \cdots + a_nb_n &\geq a_1b_n + a_2b_1 + \cdots + a_nb_{n-1},\end{aligned}$$

and adding together all the expressions, we obtain

$$n\left(a_1b_1 + \cdots + a_nb_n\right) \geq \left(a_1 + \cdots + a_n\right)\left(b_1 + \cdots + b_n\right).$$

The equality holds when $a_1 = a_2 = \cdots = a_n$ or $b_1 = b_2 = \cdots = b_n$.

Exercise 1.60. Any three positive real numbers a, b and c satisfy the following inequality:

$$a^3 + b^3 + c^3 \geq a^2b + b^2c + c^2a.$$

Exercise 1.61. Any three positive real numbers a, b and c, with $abc = 1$, satisfy

$$a^3 + b^3 + c^3 + (ab)^3 + (bc)^3 + (ca)^3 \geq 2(a^2b + b^2c + c^2a).$$

Exercise 1.62. Any three positive real numbers a, b and c satisfy

$$\frac{a^2}{b^2} + \frac{b^2}{c^2} + \frac{c^2}{a^2} \geq \frac{b}{a} + \frac{c}{b} + \frac{a}{c}.$$

Exercise 1.63. Any three positive real numbers a, b and c satisfy

$$\frac{1}{a^2} + \frac{1}{b^2} + \frac{1}{c^2} \geq \frac{a+b+c}{abc}.$$

Exercise 1.64. If a, b and c are the lengths of the sides of a triangle, prove that

$$\frac{a}{b+c-a} + \frac{b}{c+a-b} + \frac{c}{a+b-c} \geq 3.$$

Exercise 1.65. If $a_1, a_2, \ldots, a_n \in \mathbb{R}^+$ and $s = a_1 + a_2 + \cdots + a_n$, then

$$\frac{a_1}{s-a_1} + \frac{a_2}{s-a_2} + \cdots + \frac{a_n}{s-a_n} \geq \frac{n}{n-1}.$$

Exercise 1.66. If $a_1, a_2, \ldots, a_n \in \mathbb{R}^+$ and $s = a_1 + a_2 + \cdots + a_n$, then

$$\frac{s}{s-a_1} + \frac{s}{s-a_2} + \cdots + \frac{s}{s-a_n} \geq \frac{n^2}{n-1}.$$

Exercise 1.67. If $a_1, a_2, \ldots, a_n \in \mathbb{R}^+$ and $a_1 + a_2 + \cdots + a_n = 1$, then

$$\frac{a_1}{2-a_1} + \frac{a_2}{2-a_2} + \cdots + \frac{a_n}{2-a_n} \geq \frac{n}{2n-1}.$$

Exercise 1.68. (Quadratic mean-arithmetic mean inequality) Let $x_1, \ldots, x_n \in \mathbb{R}^+$, then

$$\sqrt{\frac{x_1^2 + x_2^2 + \cdots + x_n^2}{n}} \geq \frac{x_1 + x_2 + \cdots + x_n}{n}.$$

Exercise 1.69. For positive real numbers a, b, c such that $a+b+c = 1$, prove that

$$ab + bc + ca \leq \frac{1}{3}.$$

Exercise 1.70. (Harmonic, geometric and arithmetic mean) Let $x_1, \ldots, x_n \in \mathbb{R}^+$, prove that

$$\frac{n}{\frac{1}{x_1} + \frac{1}{x_2} + \cdots + \frac{1}{x_n}} \leq \sqrt[n]{x_1 x_2 \cdots x_n} \leq \frac{x_1 + x_2 + \cdots + x_n}{n}.$$

And the equalities hold if and only if $x_1 = x_2 = \cdots = x_n$.

Exercise 1.71. Let $a_1, a_2, \ldots, a_n$ be positive numbers with $a_1 a_2 \cdots a_n = 1$. Prove that

$$a_1^{n-1} + a_2^{n-1} + \cdots + a_n^{n-1} \geq \frac{1}{a_1} + \frac{1}{a_2} + \cdots + \frac{1}{a_n}.$$

Exercise 1.72. (China, 1989) Let $a_1, a_2, \ldots, a_n$ be positive numbers such that $a_1 + a_2 + \cdots + a_n = 1$. Prove that

$$\frac{a_1}{\sqrt{1-a_1}} + \cdots + \frac{a_n}{\sqrt{1-a_n}} \geq \frac{1}{\sqrt{n-1}}(\sqrt{a_1} + \cdots + \sqrt{a_n}).$$

Exercise 1.73. Let a, b and c be positive numbers such that $a+b+c=1$. Prove that

(i) $\sqrt{4a+1} + \sqrt{4b+1} + \sqrt{4c+1} < 5$,

(ii) $\sqrt{4a+1} + \sqrt{4b+1} + \sqrt{4c+1} \leq \sqrt{21}$.

Exercise 1.74. Let $a, b, c, d \in \mathbb{R}^+$ with $ab + bc + cd + da = 1$, prove that

$$\frac{a^3}{b+c+d} + \frac{b^3}{a+c+d} + \frac{c^3}{a+b+d} + \frac{d^3}{a+b+c} \geq \frac{1}{3}.$$

Exercise 1.75. Let a, b, c be positive numbers with $abc = 1$, prove that

$$\frac{a}{b} + \frac{b}{c} + \frac{c}{a} \geq a + b + c.$$

Exercise 1.76. Let $x_1, x_2, \ldots, x_n$ $(n > 2)$ be real numbers such that the sum of any $n-1$ of them is greater than the element left out of the sum. Set $s = \sum_{k=1}^{n} x_k$. Prove that

$$\sum_{k=1}^{n} \frac{x_k^2}{s - 2x_k} \geq \frac{s}{n-2}.$$

1.5 Convex functions

A function $f : [a,b] \to \mathbb{R}$ is called **convex** in the interval $I = [a,b]$ if for any $t \in [0,1]$ and for all $a \leq x < y \leq b$, the following inequality holds:

$$f(ty + (1-t)x) \leq tf(y) + (1-t)f(x). \tag{1.4}$$

Geometrically, the inequality in the definition means that the graph of f between x and y is below the segment which joins the points $(x, f(x))$ and $(y, f(y))$.

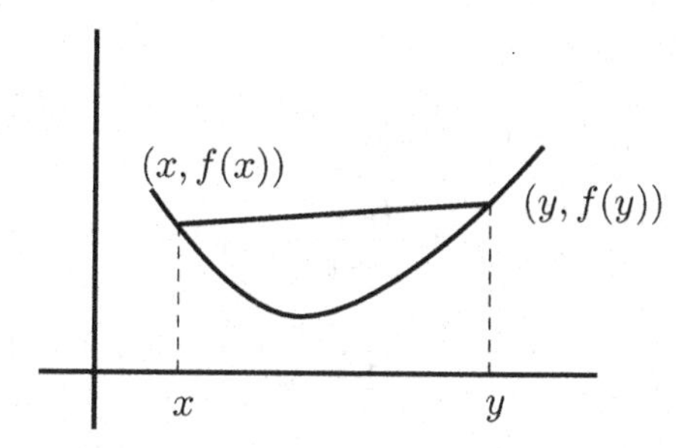

In fact, the equation of the line joining the points $(x, f(x))$ and $(y, f(y))$ is expressed as

$$L(s) = f(x) + \frac{f(y) - f(x)}{y - x}(s - x).$$

Then, evaluating at the point $s = ty + (1 - t)x$, we get

$$\begin{aligned} L(ty + (1-t)x) &= f(x) + \frac{f(y) - f(x)}{y - x}(t(y - x)) = f(x) + t(f(y) - f(x)) \\ &= tf(y) + (1-t)f(x). \end{aligned}$$

Hence, Inequality (1.4) is equivalent to

$$f(ty + (1-t)x) \leq L(ty + (1-t)x).$$

Proposition 1.5.1. (1) *If f is convex in the interval $[a, b]$, then it is convex in any subinterval $[x, y] \subset [a, b]$.*

(2) *If f is convex in $[a, b]$, then for any x, $y \in [a, b]$, we have that*

$$f\left(\frac{x + y}{2}\right) \leq \frac{1}{2}(f(x) + f(y)). \tag{1.5}$$

(3) **(Jensen's inequality)** *If f is convex in $[a, b]$, then for any $t_1, \ldots, t_n \in [0, 1]$, with $\sum_{i=1}^{n} t_i = 1$, and for $x_1, \ldots, x_n \in [a, b]$, we can deduce that*

$$f(t_1 x_1 + \cdots + t_n x_n) \leq t_1 f(x_1) + \cdots + t_n f(x_n).$$

(4) *In particular, for $x_1, \ldots, x_n \in [a, b]$, we can establish that*

$$f\left(\frac{x_1 + \cdots + x_n}{n}\right) \leq \frac{1}{n}\left(f(x_1) + \cdots + f(x_n)\right).$$

Proof. (1) We leave the proof as an exercise for the reader.

(2) It is sufficient to choose $t = \frac{1}{2}$ in (1.4).

(3) We have

$$f(t_1x_1 + \cdots + t_nx_n) = f((1-t_n)(\frac{t_1}{1-t_n}x_1 + \cdots + \frac{t_{n-1}}{1-t_n}x_{n-1}) + t_nx_n)$$

$$\leq (1-t_n) f\left(\frac{t_1}{1-t_n}x_1 + \cdots + \frac{t_{n-1}}{1-t_n}x_{n-1}\right) + t_nf(x_n), \text{ by convexity}$$

$$\leq (1-t_n)\left\{\frac{t_1}{1-t_n}f(x_1) + \cdots + \frac{t_{n-1}}{1-t_n}f(x_{n-1})\right\} + t_nf(x_n), \text{ by induction}$$

$$= t_1f(x_1) + \cdots + t_nf(x_n).$$

(4) We only need to apply (3) using $t_1 = t_2 = \cdots = t_n = \frac{1}{n}$. □

Observations 1.5.2. (i) *We can see that* (4) *holds true only under the assumption that* f *satisfies the relation* $f\left(\frac{x+y}{2}\right) \leq \frac{f(x)+f(y)}{2}$ *for any* $x,\, y \in [a, b]$.

(ii) *We can observe that* (3) *is true for* $t_1, \ldots, t_n \in [0, 1]$ *rational numbers, only under the condition that* f *satisfies the relation* $f\left(\frac{x+y}{2}\right) \leq \frac{f(x)+f(y)}{2}$ *for any* $x,\, y \in [a, b]$.

We will prove (i) using induction. Let us call P_n the assertion

$$f\left(\frac{x_1 + \cdots + x_n}{n}\right) \leq \frac{1}{n}(f(x_1) + \cdots + f(x_n))$$

for $x_1, \ldots, x_n \in [a, b]$. It is clear that P_1 and P_2 are true.

Now, we will show that $P_n \Rightarrow P_{n-1}$.

Let $x_1, \ldots, x_n \in [a, b]$ and let $y = \frac{x_1+\cdots+x_{n-1}}{n-1}$. Since P_n is true, we can establish that

$$f\left(\frac{x_1 + \cdots + x_{n-1} + y}{n}\right) \leq \frac{1}{n}f(x_1) + \cdots + \frac{1}{n}f(x_{n-1}) + \frac{1}{n}f(y).$$

But the left side is $f(y)$, therefore $n \cdot f(y) \leq f(x_1) + \cdots + f(x_{n-1}) + f(y)$, and

$$f(y) \leq \frac{1}{n-1}(f(x_1) + \cdots + f(x_{n-1})).$$

Finally, we can observe that $P_n \Rightarrow P_{2n}$.

Let $D = f\left(\frac{x_1+\cdots+x_n+x_{n+1}+\cdots+x_{2n}}{2n}\right) = f\left(\frac{u+v}{2}\right)$, where $u = \frac{x_1+\cdots+x_n}{n}$ and $v = \frac{x_{n+1}+\cdots+x_{2n}}{n}$.

Since $f\left(\frac{u+v}{2}\right) \leq \frac{1}{2}(f(u) + f(v))$, we have that

$$D \leq \frac{1}{2}(f(u) + f(v)) = \frac{1}{2}\left(f\left(\frac{x_1 + \cdots + x_n}{n}\right) + f\left(\frac{x_{n+1} + \cdots + x_{2n}}{n}\right)\right)$$

$$\leq \frac{1}{2n}(f(x_1) + \cdots + f(x_n) + f(x_{n+1}) + \cdots + f(x_{2n})),$$

where we have used twice the statement that P_n is true.

To prove (ii), our starting point will be the assertion that $f\left(\frac{x_1+\cdots+x_n}{n}\right) \leq \frac{1}{n}(f(x_1)+\cdots+f(x_n))$ for $x_1, \ldots, x_n \in [a,b]$ and $n \in \mathbb{N}$.

Let $t_1 = \frac{r_1}{s_1}, \ldots, t_n = \frac{r_n}{s_n}$ be rational numbers in $[0,1]$ with $\sum_{i=1}^n t_i = 1$. If m is the least common multiple of the s_i's, then $t_i = \frac{p_i}{m}$ with $p_i \in \mathbb{N}$ and $\sum_{i=1}^n p_i = m$, hence

$$
\begin{aligned}
f(t_1x_1+\cdots+t_nx_n) &= f\left(\frac{p_1}{m}x_1+\cdots+\frac{p_n}{m}x_n\right) \\
&= f\left(\frac{1}{m}\left[\underbrace{(x_1+\cdots+x_1)}_{p_1-\text{ terms}}+\cdots+\underbrace{(x_n+\cdots+x_n)}_{p_n-\text{ terms}}\right]\right) \\
&\leq \frac{1}{m}\left[\underbrace{(f(x_1)+\cdots+f(x_1))}_{p_1-\text{ terms}}+\cdots+\underbrace{(f(x_n)+\cdots+f(x_n))}_{p_n-\text{ terms}}\right] \\
&= \frac{p_1}{m}f(x_1)+\cdots+\frac{p_n}{m}f(x_n) \\
&= t_1f(x_1)+\cdots+t_nf(x_n).
\end{aligned}
$$

Observation 1.5.3. *If $f:[a,b]\to\mathbb{R}$ is a continuous*[2] *function on $[a,b]$ and satisfies hypothesis (2) of the proposition, then f is convex.*

We have seen that if f satisfies (2), then

$$f(qx+(1-q)y) \leq qf(x)+(1-q)f(y)$$

for any $x, y \in [a,b]$ and $q \in [0,1]$ rational number. Since any real number t can be approximated by a sequence of rational numbers q_n, and if these q_n belong to $[0,1]$, we can deduce that

$$f(q_nx+(1-q_n)y) \leq q_nf(x)+(1-q_n)f(y).$$

Now, by using the continuity of f and taking the limit, we get

$$f(tx+(1-t)y) \leq tf(x)+(1-t)f(y).$$

We say that a function $f:[a,b]\to\mathbb{R}$ is **concave** if $-f$ is convex.

[2] A function $f:[a,b]\to\mathbb{R}$ is continuous at a point $c\in[a,b]$ if $\lim_{x\to c} f(x) = f(c)$, and f is continuous on $[a,b]$ if it is continous in every point of the interval. Equivalently, f is continuous at c if for every sequence of points $\{c_n\}$ that converges to c, the sequence $\{f(c_n)\}$ converges to $f(c)$.

Observation 1.5.4. *A function $f : [a, b] \to \mathbb{R}$ is concave if and only if*

$$f(ty + (1-t)x) \geq tf(y) + (1-t)f(x) \text{ for } 0 \leq t \leq 1 \text{ and } a \leq x < y \leq b.$$

Now, we will consider some criteria to decide whether a function is convex.

Criterion 1.5.5. *A function $f : [a, b] \to \mathbb{R}$ is convex if and only if the set $\{(x, y) | a \leq x \leq b,\ f(x) \leq y\}$ is convex.*[3]

Proof. Suppose that f is convex and let $A = (x_1, y_1)$ and $B = (x_2, y_2)$ be two points in the set $U = \{(x, y) \,|\, a \leq x \leq b, f(x) \leq y\}$. To prove that $tB + (1-t)A = (tx_2 + (1-t)x_1, ty_2 + (1-t)y_1)$ belongs to U, it is sufficient to demonstrate that $a \leq tx_2 + (1-t)x_1 \leq b$ and $f(tx_2 + (1-t)x_1) \leq ty_2 + (1-t)y_1$. The first condition follows immediately since x_1 and x_2 belong to $[a, b]$.

As for the second condition, since f is convex, it follows that

$$f(tx_2 + (1-t)x_1) \leq tf(x_2) + (1-t)f(x_1).$$

Moreover, since $f(x_2) \leq y_2$ and $f(x_1) \leq y_1$, we can deduce that

$$f(tx_2 + (1-t)x_1) \leq ty_2 + (1-t)y_1.$$

Conversely, we will observe that f is convex if U is convex.

Let x_1, $x_2 \in [a, b]$ and let us consider $A = (x_1, f(x_1))$ and $B = (x_2, f(x_2))$. Clearly A and B belong to U, and since U is convex, the segment that joins them belongs to U, that is, the points of the form $tB + (1-t)A$ for $t \in [0, 1]$. Thus,

$$(tx_2 + (1-t)x_1, tf(x_2) + (1-t)f(x_1)) \in U,$$

but this implies that $f(tx_2 + (1-t)x_1) \leq tf(x_2) + (1-t)f(x_1)$. Hence f is convex. □

Criterion 1.5.6. *A function $f : [a, b] \to \mathbb{R}$ is convex if and only if, for each $x_0 \in [a, b]$, the function $P(x) = \frac{f(x)-f(x_0)}{x-x_0}$ is non-decreasing for $x \neq x_0$.*

Proof. Suppose that f is convex. To prove that $P(x)$ is non-decreasing, we take $x < y$ and then we show that $P(x) \leq P(y)$. One of the following three situations can arise: $x_0 < x < y$, $x < x_0 < y$ or $x < y < x_0$. Let us consider the first of these

[3] A subset C of the plane is convex if for any pair of points A, B in C, the segment determined by these points belongs entirely to C. Since the segment between A and B is the set of points of the form $tB + (1-t)A$, with $0 \leq t \leq 1$, the condition is that any point described by this expression belongs to C.

cases and then the other two can be proved in a similar way. First note that

$$\begin{aligned}
P(x) \leq P(y) \quad &\Leftrightarrow \quad \frac{f(x)-f(x_0)}{x-x_0} \leq \frac{f(y)-f(x_0)}{y-x_0} \\
&\Leftrightarrow \quad (f(x)-f(x_0))(y-x_0) \leq (f(y)-f(x_0))(x-x_0) \\
&\Leftrightarrow \quad f(x)(y-x_0) \leq f(y)(x-x_0) + f(x_0)(y-x) \\
&\Leftrightarrow \quad f(x) \leq f(y)\frac{x-x_0}{y-x_0} + f(x_0)\frac{y-x}{y-x_0} \\
&\Leftrightarrow \quad f\left(\frac{x-x_0}{y-x_0}y + \frac{y-x}{y-x_0}x_0\right) \leq f(y)\frac{x-x_0}{y-x_0} + f(x_0)\frac{y-x}{y-x_0}.
\end{aligned}$$

The result follows immediately. □

Criterion 1.5.7. *If the function $f : [a,b] \to \mathbb{R}$ is differentiable*[4] *with a non-decreasing derivative, then f is convex. In particular, if f is twice differentiable and $f''(x) \geq 0$, then the function is convex.*

Proof. It is clear that $f''(x) \geq 0$, for $x \in [a,b]$, implies that $f'(x)$ is non-decreasing. We see that if $f'(x)$ is non-decreasing, the function is convex.

Let $x = tb + (1-t)a$ be a point on $[a,b]$. Recalling the mean value theorem,[5] we know there exist $c \in (a,x)$ and $d \in (x,b)$ such that

$$\begin{aligned}
f(x) - f(a) &= (x-a)f'(c) = t(b-a)f'(c), \\
f(b) - f(x) &= (b-x)f'(d) = (1-t)(b-a)f'(d).
\end{aligned}$$

Then, since $f'(x)$ is non-decreasing, we can deduce that

$$(1-t)\left(f(x)-f(a)\right) = t(1-t)(b-a)f'(c) \leq t(1-t)(b-a)f'(d) = t(f(b)-f(x)).$$

After rearranging terms we get

$$f(x) \leq tf(b) + (1-t)f(a).$$ □

Let us present one **geometric interpretation of convexity** (and concavity).

Let x, y, z be points in the interval $[a,b]$ with $x < y < z$. If the vertices of the triangle XYZ have coordinates $X = (x, f(x))$, $Y = (y, f(y))$, $Z = (z, f(z))$, then the area of the triangle is given by

$$\Delta = \frac{1}{2}\det A, \text{ where } A = \begin{pmatrix} 1 & x & f(x) \\ 1 & y & f(y) \\ 1 & z & f(z) \end{pmatrix}.$$

[4] A function $f : [a,b] \to \mathbb{R}$ is differentiable in a point $c \in [a,b]$ if the function $f'(c) = \lim\limits_{x \to c} \frac{f(x)-f(c)}{x-c}$ exists and f is differentiable in $A \subset [a,b]$ if it is differentiable in every point of A.

[5] **Mean value theorem.** For a continuous function $f : [a,b] \to \mathbb{R}$, which is differentiable in (a,b), there exists a number $x \in (a,b)$ such that $f'(x)(b-a) = f(b) - f(a)$. See [21, page 169].

The area can be positive or negative, this will depend on whether the triangle XYZ is positively oriented (anticlockwise oriented) or negatively oriented. For a convex function, we have that $\Delta > 0$ and for a concave function, $\Delta < 0$, as shown in the following graphs.

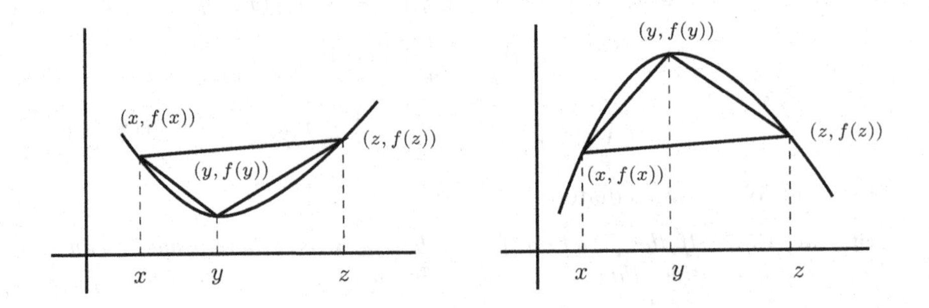

In fact,

$$\begin{aligned}\Delta > 0 &\Leftrightarrow \det A > 0\\ &\Leftrightarrow (z-y)f(x) - (z-x)f(y) + (y-x)f(z) > 0\\ &\Leftrightarrow f(y) < \frac{z-y}{z-x}f(x) + \frac{y-x}{z-x}f(z).\end{aligned}$$

If we take $t = \frac{y-x}{z-x}$, we have $0 < t < 1$, $1 - t = \frac{z-y}{z-x}$, $y = tz + (1-t)x$ and $f(tz + (1-t)x) < tf(z) + (1-t)f(x)$.

Now, let us introduce several examples where convex functions are used to establish inequalities.

Example 1.5.8. *The function $f(x) = x^n$, $n \geq 1$, is convex in $\mathbb{R}^+$ and the function $f(x) = x^n$, with n even, is also convex in $\mathbb{R}$.*

This follows from the fact that $f''(x) = n(n-1)x^{n-2} \geq 0$ in each case.

As an application of this we get the following.

(i) Since $\left(\frac{a+b}{2}\right)^2 \leq \frac{a^2+b^2}{2}$, we can deduce that $\frac{a+b}{2} \leq \sqrt{\frac{a^2+b^2}{2}}$, which is the inequality between the arithmetic mean and the quadratic mean.

(ii) Since $\left(\frac{a+b}{2}\right)^n \leq \frac{a^n+b^n}{2}$, we can deduce that $a^n + b^n \geq \dfrac{1}{2^{n-1}}$, for a and b positive numbers such that $a + b = 1$.

(iii) If a and b are positive numbers, $\left(1 + \frac{a}{b}\right)^n + \left(1 + \frac{b}{a}\right)^n \geq 2^{n+1}$. This follows from

$$2^n = f(2) \leq f\left(\frac{\frac{a+b}{a} + \frac{a+b}{b}}{2}\right) \leq \frac{1}{2}\left[f\left(1+\frac{a}{b}\right) + f\left(1+\frac{b}{a}\right)\right]$$
$$= \frac{1}{2}\left[\left(1+\frac{a}{b}\right)^n + \left(1+\frac{b}{a}\right)^n\right].$$

Example 1.5.9. *The exponential function $f(x) = e^x$ is convex in $\mathbb{R}$, since $f''(x) = e^x > 0$, for every $x \in \mathbb{R}$.*

Let us observe several ways in which this property can be used.

(i) **(Weighted AM-GM inequality)** If $x_1, \ldots, x_n, t_1, \ldots, t_n$ are positive numbers and $\sum_{i=1}^n t_i = 1$, then

$$x_1^{t_1} \cdots x_n^{t_n} \leq t_1x_1 + \cdots + t_nx_n.$$

In fact, since $x_i^{t_i} = e^{t_i \log x_i}$ and e^x is convex, we can deduce that

$$x_1^{t_1} \cdots x_n^{t_n} = e^{t_1 \log x_1} \cdots e^{t_n \log x_n} = e^{t_1 \log x_1 + \cdots + t_n \log x_n}$$
$$\leq t_1 e^{\log x_1} + \cdots + t_n e^{\log x_n} = t_1x_1 + \cdots + t_nx_n.$$

In particular, if we take $t_i = \frac{1}{n}$, for $1 \leq i \leq n$, we can produce another proof of the inequality between the arithmetic mean and the geometric mean for n numbers.

(ii) **(Young's inequality)** Let x, y be positive real numbers. If $a, b > 0$ satisfy the condition $\frac{1}{a} + \frac{1}{b} = 1$, then $xy \leq \frac{1}{a}x^a + \frac{1}{b}y^b$.

We only need to apply part (i) as follows:

$$xy = (x^a)^{\frac{1}{a}} \left(y^b\right)^{\frac{1}{b}} \leq \frac{1}{a}x^a + \frac{1}{b}y^b.$$

(iii) **(Hölder's inequality)** Let $x_1, x_2, \ldots, x_n, y_1, y_2, \ldots, y_n$ be positive numbers and $a, b > 0$ such that $\frac{1}{a} + \frac{1}{b} = 1$, then

$$\sum_{i=1}^n x_iy_i \leq \left(\sum_{i=1}^n x_i^a\right)^{1/a} \left(\sum_{i=1}^n y_i^b\right)^{1/b}.$$

Let us first assume that $\sum_{i=1}^n x_i^a = \sum_{i=1}^n y_i^b = 1$.

Using part (ii), $x_iy_i \leq \frac{1}{a}x_i^a + \frac{1}{b}y_i^b$, then

$$\sum_{i=1}^n x_iy_i \leq \frac{1}{a}\sum_{i=1}^n x_i^a + \frac{1}{b}\sum_{i=1}^n y_i^b = \frac{1}{a} + \frac{1}{b} = 1.$$

Now, suppose that $\sum_{i=1}^n x_i^a = A$ and $\sum_{i=1}^n y_i^b = B$. Let us take $x_i' = \frac{x_i}{A^{1/a}}$ and $y_i' = \frac{y_i}{B^{1/b}}$. Since

$$\sum_{i=1}^n (x_i')^a = \frac{\sum_{i=1}^n x_i^a}{A} = 1 \quad \text{and} \quad \sum_{i=1}^n (y_i')^b = \frac{\sum_{i=1}^n y_i^b}{B} = 1,$$

we can deduce that

$$1 \geq \sum_{i=1}^n x_i' y_i' = \sum_{i=1}^n \frac{x_i y_i}{A^{1/a} B^{1/b}} = \frac{1}{A^{1/a} B^{1/b}} \sum_{i=1}^n x_i y_i.$$

Therefore, $\sum_{i=1}^n x_i y_i \leq A^{1/a} B^{1/b}$.

If we choose $a = b = 2$, we get the Cauchy-Schwarz inequality.

Let us introduce a consequence of Hölder's inequality, which is a generalization of the triangle inequality.

Example 1.5.10 (Minkowski's inequality). *Let $a_1, a_2, \ldots, a_n, b_1, b_2, \ldots, b_n$ be positive numbers and $p > 1$, then*

$$\left(\sum_{k=1}^n (a_k + b_k)^p\right)^{\frac{1}{p}} \leq \left(\sum_{k=1}^n (a_k)^p\right)^{\frac{1}{p}} + \left(\sum_{k=1}^n (b_k)^p\right)^{\frac{1}{p}}.$$

We note that

$$(a_k + b_k)^p = a_k (a_k + b_k)^{p-1} + b_k (a_k + b_k)^{p-1},$$

so that

$$\sum_{k=1}^n (a_k + b_k)^p = \sum_{k=1}^n a_k (a_k + b_k)^{p-1} + \sum_{k=1}^n b_k (a_k + b_k)^{p-1}. \tag{1.6}$$

We apply Hölder's inequality to each term of the sum on the right-hand side of (1.6), with q such that $\frac{1}{p} + \frac{1}{q} = 1$, to get

$$\sum_{k=1}^n a_k (a_k + b_k)^{p-1} \leq \left(\sum_{k=1}^n (a_k)^p\right)^{\frac{1}{p}} \left(\sum_{k=1}^n (a_k + b_k)^{q(p-1)}\right)^{\frac{1}{q}},$$

$$\sum_{k=1}^n b_k (a_k + b_k)^{p-1} \leq \left(\sum_{k=1}^n (b_k)^p\right)^{\frac{1}{p}} \left(\sum_{k=1}^n (a_k + b_k)^{q(p-1)}\right)^{\frac{1}{q}}.$$

Putting these inequalities into (1.6), and noting that $q(p-1) = p$, yields the required inequality. Note that Minkowski's inequality is an equality if we allow $p = 1$. For $0 < p < 1$, the inequality is reversed.

Example 1.5.11. (Short list IMO, 1998) *If $r_1, \ldots, r_n$ are real numbers greater than 1, prove that*

$$\frac{1}{1+r_1}+\cdots+\frac{1}{1+r_n} \geq \frac{n}{\sqrt[n]{r_1\cdots r_n}+1}.$$

First note that the function $f(x)=\frac{1}{1+e^x}$ is convex for $\mathbb{R}^+$, since $f'(x)=\frac{-e^x}{(1+e^x)^2}$ and $f''(x)=\frac{e^x(e^x-1)}{(e^x+1)^3}\geq 0$ for $x>0$.

Now, if $r_i>1$, then $r_i=e^{x_i}$ for some $x_i>0$. Since $f(x)=\frac{1}{1+e^x}$ is convex, we can establish that

$$\frac{1}{e^{\left(\frac{x_1+\cdots+x_n}{n}\right)}+1} \leq \frac{1}{n}\left(\frac{1}{1+e^{x_1}}+\cdots+\frac{1}{1+e^{x_n}}\right),$$

hence

$$\frac{n}{\sqrt[n]{r_1\cdots r_n}+1} \leq \frac{1}{1+r_1}+\cdots+\frac{1}{1+r_n}.$$

Example 1.5.12. (China, 1989) *Prove that for any n real positive numbers $x_1, \ldots, x_n$ such that $\sum_{i=1}^n x_i = 1$, we have*

$$\sum_{i=1}^n \frac{x_i}{\sqrt{1-x_i}} \geq \frac{\sum_{i=1}^n \sqrt{x_i}}{\sqrt{n-1}}.$$

We will use the fact that the function $f(x)=\frac{x}{\sqrt{1-x}}$ is convex in $(0,1)$, since $f''(x)>0$,

$$\frac{1}{n}\sum_{i=1}^n \frac{x_i}{\sqrt{1-x_i}} = \frac{1}{n}\sum_{i=1}^n f(x_i) \geq f\left(\sum_{i=1}^n \frac{1}{n}x_i\right) = f\left(\frac{1}{n}\right) = \frac{1}{\sqrt{n}\sqrt{n-1}},$$

hence

$$\sum_{i=1}^n \frac{x_i}{\sqrt{1-x_i}} \geq \frac{\sqrt{n}}{\sqrt{n-1}}.$$

It is left to prove that $\sum_{i=1}^n \sqrt{x_i} \leq \sqrt{n}$, but this follows from the Cauchy-Schwarz inequality, $\sum_{i=1}^n \sqrt{x_i} \leq \sqrt{\sum_{i=1}^n x_i}\sqrt{\sum_{i=1}^n 1} = \sqrt{n}$.

Example 1.5.13. (Hungary–Israel, 1999) *Let k and l be two given positive integers, and let a_{ij}, $1\leq i\leq k$ and $1\leq j\leq l$, be kl given positive numbers. Prove that if $q\geq p>0$, then*

$$\left(\sum_{j=1}^l\left(\sum_{i=1}^k a_{ij}^p\right)^{\frac{q}{p}}\right)^{\frac{1}{q}} \leq \left(\sum_{i=1}^k\left(\sum_{j=1}^l a_{ij}^q\right)^{\frac{p}{q}}\right)^{\frac{1}{p}}.$$

Define $b_j = \sum_{i=1}^k a_{ij}^p$ for $j = 1, 2, \ldots, l$, and denote the left-hand side of the required inequality by L and the right-hand side by R. Then

$$\begin{aligned} L^q &= \sum_{j=1}^{l} b_j^{\frac{q}{p}} \\ &= \sum_{j=1}^{l} \left(b_j^{\frac{q-p}{p}} \left(\sum_{i=1}^{k} a_{ij}^p \right) \right) \\ &= \sum_{i=1}^{k} \left(\sum_{j=1}^{l} b_j^{\frac{q-p}{p}} a_{ij}^p \right). \end{aligned}$$

Using Hölder's inequality we obtain

$$\begin{aligned} L^q &\leq \sum_{i=1}^{k} \left[\left(\sum_{j=1}^{l} \left(b_j^{\frac{q-p}{p}} \right)^{\frac{q}{q-p}} \right)^{\frac{q-p}{q}} \left(\sum_{j=1}^{l} (a_{ij}^p)^{\frac{q}{p}} \right)^{\frac{p}{q}} \right] \\ &= \sum_{i=1}^{k} \left[\left(\sum_{j=1}^{l} b_j^{\frac{q}{p}} \right)^{\frac{q-p}{q}} \left(\sum_{j=1}^{l} a_{ij}^q \right)^{\frac{p}{q}} \right] \\ &= \left(\sum_{j=1}^{l} b_j^{\frac{q}{p}} \right)^{\frac{q-p}{q}} \cdot \left[\sum_{i=1}^{k} \left(\sum_{j=1}^{l} a_{ij}^q \right)^{\frac{p}{q}} \right] = L^{q-p} R^p. \end{aligned}$$

The inequality $L \leq R$ follows by dividing both sides of $L^q \leq L^{q-p} R^p$ by L^{q-p} and taking the p-th root.

Exercise 1.77. (i) For $a, b \in \mathbb{R}^+$, with $a + b = 1$, prove that

$$\left(a + \frac{1}{a} \right)^2 + \left(b + \frac{1}{b} \right)^2 \geq \frac{25}{2}.$$

(ii) For $a, b, c \in \mathbb{R}^+$, with $a + b + c = 1$, prove that

$$\left(a + \frac{1}{a} \right)^2 + \left(b + \frac{1}{b} \right)^2 + \left(c + \frac{1}{c} \right)^2 \geq \frac{100}{3}.$$

Exercise 1.78. For $0 \leq a, b, c \leq 1$, prove that

$$\frac{a}{b+c+1} + \frac{b}{c+a+1} + \frac{c}{a+b+1} + (1-a)(1-b)(1-c) \leq 1.$$

Exercise 1.79. (Russia, 2000) For real numbers x, y such that $0 \leq x, y \leq 1$, prove that

$$\frac{1}{\sqrt{1+x^2}} + \frac{1}{\sqrt{1+y^2}} \leq \frac{2}{\sqrt{1+xy}}.$$

Exercise 1.80. Prove that the function $f(x) = \sin x$ is concave in the interval $[0, \pi]$. Use this to verify that the angles A, B, C of a triangle satisfy $\sin A + \sin B + \sin C \leq \frac{3}{2}\sqrt{3}$.

Exercise 1.81. If A, B, C, D are angles belonging to the interval $[0, \pi]$, then

(i) $\sin A \sin B \leq \sin^2\left(\frac{A+B}{2}\right)$ and the equality holds if and only if $A = B$,

(ii) $\sin A \sin B \sin C \sin D \leq \sin^4\left(\frac{A+B+C+D}{4}\right)$,

(iii) $\sin A \sin B \sin C \leq \sin^3\left(\frac{A+B+C}{3}\right)$,

Moreover, if A, B, C are the internal angles of a triangle, then

(iv) $\sin A \sin B \sin C \leq \frac{3}{8}\sqrt{3}$,

(v) $\sin\frac{A}{2}\sin\frac{B}{2}\sin\frac{C}{2} \leq \frac{1}{8}$,

(vi) $\sin A + \sin B + \sin C = 4\cos\frac{A}{2}\cos\frac{B}{2}\cos\frac{C}{2}$.

Exercise 1.82. (Bernoulli's inequality)

(i) For any real number $x > -1$ and for every positive integer n, we have $(1+x)^n \geq 1 + nx$.

(ii) Use this inequality to provide another proof of the AM-GM inequality.

Exercise 1.83. (Schür's inequality) If x, y, z are positive real numbers and n is a positive integer, we have

$$x^n(x-y)(x-z) + y^n(y-z)(y-x) + z^n(z-x)(z-y) \geq 0.$$

For the case $n = 1$, the inequality can take one of the following forms:

(a) $x^3 + y^3 + z^3 + 3xyz \geq xy(x+y) + yz(y+z) + zx(z+x)$.

(b) $xyz \geq (x+y-z)(y+z-x)(z+x-y)$.

(c) If $x + y + z = 1$, $9xyz + 1 \geq 4(xy + yz + zx)$.

Exercise 1.84. (Canada, 1992) For any three non-negative real numbers x, y and z we have

$$x(x-z)^2 + y(y-z)^2 \geq (x-z)(y-z)(x+y-z).$$

Exercise 1.85. If a, b, c are positive real numbers, prove that

$$\frac{a}{(b+c)^2} + \frac{b}{(c+a)^2} + \frac{c}{(a+b)^2} \geq \frac{9}{4(a+b+c)}.$$

Exercise 1.86. Let a, b and c be positive real numbers, prove that

$$1 + \frac{3}{ab+bc+ca} \geq \frac{6}{a+b+c}.$$

Moreover, if $abc = 1$, prove that

$$1 + \frac{3}{a+b+c} \geq \frac{6}{ab+bc+ca}.$$

Exercise 1.87. (Power mean inequality) Let $x_1, x_2, \ldots, x_n$ be positive real numbers and let $t_1, t_2, \ldots, t_n$ be positive real numbers adding up to 1. Let r and s be two nonzero real numbers such that $r > s$. Prove that

$$(t_1 x_1^r + \cdots + t_n x_n^r)^{\frac{1}{r}} \geq (t_1 x_1^s + \cdots + t_n x_n^s)^{\frac{1}{s}}$$

with equality if and only if $x_1 = x_2 = \cdots = x_n$.

Exercise 1.88. (Two extensions of Hölder's inequality) Let $x_1, x_2, \ldots, x_n$, $y_1, y_2, \ldots, y_n$, $z_1, z_2, \ldots, z_n$ be positive real numbers.

(i) If a, b, c are positive real numbers such that $\frac{1}{a} + \frac{1}{b} = \frac{1}{c}$, then

$$\left\{\sum_{i=1}^{n} (x_i y_i)^c\right\}^{\frac{1}{c}} \leq \left\{\sum_{i=1}^{n} {x_i}^a\right\}^{\frac{1}{a}} \left\{\sum_{i=1}^{n} {y_i}^b\right\}^{\frac{1}{b}}.$$

(ii) If a, b, c are positive real numbers such that $\frac{1}{a} + \frac{1}{b} + \frac{1}{c} = 1$, then

$$\sum_{i=1}^{n} x_i y_i z_i \leq \left\{\sum_{i=1}^{n} {x_i}^a\right\}^{\frac{1}{a}} \left\{\sum_{i=1}^{n} {y_i}^b\right\}^{\frac{1}{b}} \left\{\sum_{i=1}^{n} {z_i}^c\right\}^{\frac{1}{c}}.$$

Exercise 1.89. (Popoviciu's inequality) If I is an interval and $f : I \to \mathbb{R}$ is a convex function, then for a, b, $c \in I$ the following inequality holds:

$$\frac{2}{3}\left[f\left(\frac{a+b}{2}\right) + f\left(\frac{b+c}{2}\right) + f\left(\frac{c+a}{2}\right)\right] \leq \frac{f(a)+f(b)+f(c)}{3} + f\left(\frac{a+b+c}{3}\right).$$

Exercise 1.90. Let a, b, c be non-negative real numbers. Prove that

(i) $a^2 + b^2 + c^2 + 3\sqrt[3]{a^2b^2c^2} \geq 2(ab+bc+ca)$,

(ii) $a^2 + b^2 + c^2 + 2abc + 1 \geq 2(ab+bc+ca)$.

Exercise 1.91. Let a, b, c be positive real numbers. Prove that

$$\left(\frac{b+c}{a} + \frac{c+a}{b} + \frac{a+b}{c}\right) \geq 4\left(\frac{a}{b+c} + \frac{b}{c+a} + \frac{c}{a+b}\right).$$

1.6 A helpful inequality

First, let us study two very useful algebraic identities that are deduced by considering a special factor of $a^3 + b^3 + c^3 - 3abc$.

Let P denote the cubic polynomial

$$P(x) = x^3 - (a + b + c)x^2 + (ab + bc + ca)x - abc,$$

which has a, b and c as its roots. By substituting a, b, c in the polynomial, we obtain

$$a^3 - (a + b + c)a^2 + (ab + bc + ca)a - abc = 0,$$

$$b^3 - (a + b + c)b^2 + (ab + bc + ca)b - abc = 0,$$

$$c^3 - (a + b + c)c^2 + (ab + bc + ca)c - abc = 0.$$

Adding up these three equations yields

$$a^3 + b^3 + c^3 - 3abc = (a + b + c)(a^2 + b^2 + c^2 - ab - bc - ca). \qquad (1.7)$$

It immediately follows that if $a + b + c = 0$, then $a^3 + b^3 + c^3 = 3abc$.

Note also that the expression

$$a^2 + b^2 + c^2 - ab - bc - ca$$

can also be written as

$$a^2 + b^2 + c^2 - ab - bc - ca = \frac{1}{2}[(a - b)^2 + (b - c)^2 + (c - a)^2]. \qquad (1.8)$$

In this way, we obtain another version of identity (1.7),

$$a^3 + b^3 + c^3 - 3abc = \frac{1}{2}(a + b + c)[(a - b)^2 + (b - c)^2 + (c - a)^2]. \qquad (1.9)$$

This presentation of the identity leads to a short proof of the AM-GM inequality for three variables. From (1.9) it is clear that if a, b, c are positive numbers, then $a^3 + b^3 + c^3 \geq 3abc$. Now, if x, y, z are positive numbers, taking $a = \sqrt[3]{x}$, $b = \sqrt[3]{y}$ and $c = \sqrt[3]{z}$ will lead us to

$$\frac{x + y + z}{3} \geq \sqrt[3]{xyz}$$

with equality if and only if $x = y = z$.

Note that identity (1.8) provides another proof of Exercise 1.27.

Exercise 1.92. For real numbers x, y, z, prove that

$$x^2 + y^2 + z^2 \geq |xy + yz + zx|.$$

Exercise 1.93. For positive real numbers a, b, c, prove that

$$\frac{a^2+b^2+c^2}{abc} \geq \frac{1}{a}+\frac{1}{b}+\frac{1}{c}.$$

Exercise 1.94. If x, y, z are real numbers such that $x < y < z$, prove that

$$(x-y)^3+(y-z)^3+(z-x)^3 > 0.$$

Exercise 1.95. Let a, b, c be the side lengths of a triangle. Prove that

$$\sqrt[3]{\frac{a^3+b^3+c^3+3abc}{2}} \geq \max\{a,b,c\}.$$

Exercise 1.96. (Romania, 2007) For non-negative real numbers x, y, z, prove that

$$\frac{x^3+y^3+z^3}{3} \geq xyz + \frac{3}{4}|(x-y)(y-z)(z-x)|.$$

Exercise 1.97. (UK, 2008) Find the minimum of $x^2+y^2+z^2$, where x, y, z are real numbers such that $x^3+y^3+z^3-3xyz=1$.

A very simple inequality which may be helpful for proving a large number of algebraic inequalities is the following.

Theorem 1.6.1 (A helpful inequality). *If a, b, x, y are real numbers and x, $y > 0$, then the following inequality holds:*

$$\frac{a^2}{x}+\frac{b^2}{y} \geq \frac{(a+b)^2}{x+y}. \tag{1.10}$$

Proof. The proof is quite simple. Clearing out denominators, we can express the inequality as

$$a^2y(x+y)+b^2x(x+y) \geq (a+b)^2xy,$$

which simplifies to become the obvious $(ay-bx)^2 \geq 0$. We see that the equality holds if and only if $ay = bx$, that is, if and only if $\frac{a}{x}=\frac{b}{y}$.

Another form to prove the inequality is using the Cauchy-Schwarz inequality in the following way:

$$(a+b)^2 = \left(\frac{a}{\sqrt{x}}\sqrt{x}+\frac{b}{\sqrt{y}}\sqrt{y}\right)^2 \leq \left(\frac{a^2}{x}+\frac{b^2}{y}\right)(x+y). \qquad \square$$

Using the above theorem twice, we can extend the inequality to three pairs of numbers

$$\frac{a^2}{x}+\frac{b^2}{y}+\frac{c^2}{z} \geq \frac{(a+b)^2}{x+y}+\frac{c^2}{z} \geq \frac{(a+b+c)^2}{x+y+z},$$

and a simple inductive argument shows that

$$\frac{a_1^2}{x_1} + \frac{a_2^2}{x_2} + \cdots + \frac{a_n^2}{x_n} \geq \frac{(a_1 + a_2 + \cdots + a_n)^2}{x_1 + x_2 + \cdots + x_n} \tag{1.11}$$

for all real numbers $a_1, a_2, \ldots, a_n$ and $x_1, x_2, \ldots, x_n > 0$, with equality if and only if

$$\frac{a_1}{x_1} = \frac{a_2}{x_2} = \cdots = \frac{a_n}{x_n}.$$

Inequality (1.11) is also called the Cauchy-Schwarz inequality in Engel form or Arthur Engel's Minima Principle.

As a first application of this inequality, we will present another proof of the Cauchy-Schwarz inequality. Let us write

$$a_1^2 + a_2^2 + \cdots + a_n^2 = \frac{a_1^2 b_1^2}{b_1^2} + \frac{a_2^2 b_2^2}{b_2^2} + \cdots + \frac{a_n^2 b_n^2}{b_n^2},$$

then

$$\frac{a_1^2 b_1^2}{b_1^2} + \frac{a_2^2 b_2^2}{b_2^2} + \cdots + \frac{a_n^2 b_n^2}{b_n^2} \geq \frac{(a_1 b_1 + a_2 b_2 + \cdots + a_n b_n)^2}{b_1^2 + b_2^2 + \cdots + b_n^2}.$$

Thus, we conclude that

$$(a_1^2 + a_2^2 + \cdots + a_n^2)(b_1^2 + b_2^2 + \cdots + b_n^2) \geq (a_1 b_1 + a_2 b_2 + \cdots + a_n b_n)^2$$

and the equality holds if and only if

$$\frac{a_1}{b_1} = \frac{a_2}{b_2} = \cdots = \frac{a_n}{b_n}.$$

It is worth to mention that there are other forms of the Cauchy-Schwarz inequality in Engel form.

Example 1.6.2. *Let $a_1, \ldots, a_n, b_1, \ldots, b_n$ be positive real numbers. Prove that*

(i) $\displaystyle \frac{a_1}{b_1} + \cdots + \frac{a_n}{b_n} \geq \frac{(a_1 + \cdots + a_n)^2}{a_1 b_1 + \cdots + a_n b_n}$,

(ii) $\displaystyle \frac{a_1}{b_1^2} + \cdots + \frac{a_n}{b_n^2} \geq \frac{1}{a_1 + \cdots + a_n} \left(\frac{a_1}{b_1} + \cdots + \frac{a_n}{b_n} \right)^2$.

Both inequalities are direct consequence of inequality (1.11), as we can see as follows.

(i) $\displaystyle \frac{a_1}{b_1} + \cdots + \frac{a_n}{b_n} = \frac{a_1^2}{a_1 b_1} + \cdots + \frac{a_n^2}{a_n b_n} \geq \frac{(a_1 + \cdots + a_n)^2}{a_1 b_1 + \cdots + a_n b_n}$,

(ii) $\displaystyle \frac{a_1}{b_1^2} + \cdots + \frac{a_n}{b_n^2} = \frac{\frac{a_1^2}{b_1^2}}{a_1} + \cdots + \frac{\frac{a_n^2}{b_n^2}}{a_n} \geq \frac{1}{a_1 + \cdots + a_n} \left(\frac{a_1}{b_1} + \cdots + \frac{a_n}{b_n} \right)^2$.

Example 1.6.3. (APMO, 1991) *Let $a_1, \ldots, a_n, b_1, \ldots, b_n$ be positive real numbers such that $a_1 + a_2 + \cdots + a_n = b_1 + b_2 + \cdots + b_n$. Prove that*

$$\frac{a_1^2}{a_1+b_1} + \cdots + \frac{a_n^2}{a_n+b_n} \geq \frac{1}{2}(a_1 + \cdots + a_n).$$

Observe that (1.11) implies that

$$\begin{aligned}\frac{a_1^2}{a_1+b_1} + \cdots + \frac{a_n^2}{a_n+b_n} &\geq \frac{(a_1+a_2+\cdots+a_n)^2}{a_1+a_2+\cdots+a_n+b_1+b_2+\cdots+b_n} \\ &= \frac{(a_1+a_2+\cdots+a_n)^2}{2(a_1+a_2+\cdots+a_n)} \\ &= \frac{1}{2}(a_1+a_2+\cdots+a_n).\end{aligned}$$

The following example consists of a proof of the quadratic mean-arithmetic mean inequality.

Example 1.6.4 (Quadratic mean-arithmetic mean inequality). *For positive real numbers $x_1, \ldots, x_n$, we have*

$$\sqrt{\frac{x_1^2 + x_2^2 + \cdots + x_n^2}{n}} \geq \frac{x_1 + x_2 + \cdots + x_n}{n}.$$

Observe that using (1.11) leads us to

$$\frac{x_1^2 + x_2^2 + \cdots + x_n^2}{n} \geq \frac{(x_1 + x_2 + \cdots + x_n)^2}{n^2},$$

which implies the above inequality.

In some cases the numerators are not squares, but a simple trick allows us to write them as squares, so that we can use the inequality. Our next application shows this trick and offers a shorter proof for Example 1.4.9.

Example 1.6.5. (IMO, 1995) *Let a, b, c be positive real numbers such that $abc = 1$. Prove that*

$$\frac{1}{a^3(b+c)} + \frac{1}{b^3(a+c)} + \frac{1}{c^3(a+b)} \geq \frac{3}{2}.$$

Observe that

$$\begin{aligned}\frac{1}{a^3(b+c)} + \frac{1}{b^3(c+a)} + \frac{1}{c^3(a+b)} &= \frac{\frac{1}{a^2}}{a(b+c)} + \frac{\frac{1}{b^2}}{b(c+a)} + \frac{\frac{1}{c^2}}{c(a+b)} \\ &\geq \frac{(\frac{1}{a}+\frac{1}{b}+\frac{1}{c})^2}{2(ab+bc+ca)} = \frac{ab+bc+ca}{2(abc)} \\ &\geq \frac{3\sqrt[3]{(abc)^2}}{2} = \frac{3}{2},\end{aligned}$$

where the first inequality follows from (1.11) and the second is a consequence of the *AM-GM* inequality.

As a further example of the use of inequality (1.11), we provide a simple proof of Nesbitt's inequality.

Example 1.6.6 (Nesbitt's inequality). *For $a,\ b,\ c \in \mathbb{R}^+$, we have*

$$\frac{a}{b+c} + \frac{b}{c+a} + \frac{c}{a+b} \geq \frac{3}{2}.$$

We multiply the three terms on the left-hand side of the inequality by $\frac{a}{a}$, $\frac{b}{b}$, $\frac{c}{c}$, respectively, and then we use inequality (1.11) to produce

$$\frac{a^2}{a(b+c)} + \frac{b^2}{b(c+a)} + \frac{c^2}{c(a+b)} \geq \frac{(a+b+c)^2}{2(ab+bc+ca)}.$$

From Equation (1.8) we know that $a^2 + b^2 + c^2 - ab - bc - ca \geq 0$, that is, $(a+b+c)^2 \geq 3(ab+bc+ca)$. Therefore

$$\frac{a}{b+c} + \frac{b}{c+a} + \frac{c}{a+b} \geq \frac{(a+b+c)^2}{2(ab+bc+ca)} \geq \frac{3}{2}.$$

Example 1.6.7. (Czech and Slovak Republics, 1999) *For a, b and c positive real numbers, prove the inequality*

$$\frac{a}{b+2c} + \frac{b}{c+2a} + \frac{c}{a+2b} \geq 1.$$

Observe that

$$\frac{a}{b+2c} + \frac{b}{c+2a} + \frac{c}{a+2b} = \frac{a^2}{ab+2ca} + \frac{b^2}{bc+2ab} + \frac{c^2}{ca+2bc}.$$

Then using (1.11) yields

$$\frac{a^2}{ab+2ca} + \frac{b^2}{bc+2ab} + \frac{c^2}{ca+2bc} \geq \frac{(a+b+c)^2}{3(ab+bc+ca)} \geq 1,$$

where the last inequality follows in the same way as in the previous example.

Exercise 1.98. (South Africa, 1995) For a, b, c, d positive real numbers, prove that

$$\frac{1}{a} + \frac{1}{b} + \frac{4}{c} + \frac{16}{d} \geq \frac{64}{a+b+c+d}.$$

Exercise 1.99. Let a and b be positive real numbers. Prove that

$$8(a^4 + b^4) \geq (a+b)^4.$$

Exercise 1.100. Let x, y, z be positive real numbers. Prove that

$$\frac{2}{x+y}+\frac{2}{y+z}+\frac{2}{z+x}\geq\frac{9}{x+y+z}.$$

Exercise 1.101. Let a, b, x, y, z be positive real numbers. Prove that

$$\frac{x}{ay+bz}+\frac{y}{az+bx}+\frac{z}{ax+by}\geq\frac{3}{a+b}.$$

Exercise 1.102. Let a, b, c be positive real numbers. Prove that

$$\frac{a^2+b^2}{a+b}+\frac{b^2+c^2}{b+c}+\frac{c^2+a^2}{c+a}\geq a+b+c.$$

Exercise 1.103. (i) Let x, y, z be positive real numbers. Prove that

$$\frac{x}{x+2y+3z}+\frac{y}{y+2z+3x}+\frac{z}{z+2x+3y}\geq\frac{1}{2}.$$

(ii) (Moldova, 2007) Let w, x, y, z be positive real numbers. Prove that

$$\frac{w}{x+2y+3z}+\frac{x}{y+2z+3w}+\frac{y}{z+2w+3x}+\frac{z}{w+2x+3y}\geq\frac{2}{3}.$$

Exercise 1.104. (Croatia, 2004) Let x, y, z be positive real numbers. Prove that

$$\frac{x^2}{(x+y)(x+z)}+\frac{y^2}{(y+z)(y+x)}+\frac{z^2}{(z+x)(z+y)}\geq\frac{3}{4}.$$

Exercise 1.105. For a, b, c, d positive real numbers, prove that

$$\frac{a}{b+c}+\frac{b}{c+d}+\frac{c}{d+a}+\frac{d}{a+b}\geq 2.$$

Exercise 1.106. Let a, b, c, d, e be positive real numbers. Prove that

$$\frac{a}{b+c}+\frac{b}{c+d}+\frac{c}{d+e}+\frac{d}{e+a}+\frac{e}{a+b}\geq\frac{5}{2}.$$

Exercise 1.107. (i) Prove that, for all positive real numbers a, b, c, x, y, z with $a\geq b\geq c$ and $z\geq y\geq x$, the following inequality holds:

$$\frac{a^3}{x}+\frac{b^3}{y}+\frac{c^3}{z}\geq\frac{(a+b+c)^3}{3(x+y+z)}.$$

(ii) (Belarus, 2000) Prove that, for all positive real numbers a, b, c, x, y, z, the following inequality holds:

$$\frac{a^3}{x}+\frac{b^3}{y}+\frac{c^3}{z}\geq\frac{(a+b+c)^3}{3(x+y+z)}.$$

Exercise 1.108. (Greece, 2008) For x_1, x_2, ..., x_n positive integers, prove that

$$\left(\frac{x_1^2 + x_2^2 + \cdots + x_n^2}{x_1 + x_2 + \cdots + x_n}\right)^{\frac{kn}{t}} \geq x_1 \cdot x_2 \cdot \cdots \cdot x_n,$$

where $k = \max\{x_1, x_2, \ldots, x_n\}$ and $t = \min\{x_1, x_2, \ldots, x_n\}$. Under which condition the equality holds?

1.7 The substitution strategy

Substitution is a useful strategy to solve inequality problems. Making an adequate substitution we can, for instance, change the difficult terms of the inequality a little, we can simplify expressions or we can reduce terms. In this section we give some ideas of what can be done with this strategy. As always, the best way to do that is through some examples.

One useful suggestion for problems that contain in the hypothesis an extra condition, is to use that condition to simplify the problem. In the next example we apply this technique to eliminate the denominators in order to make the problem easier to solve.

Example 1.7.1. *If a, b, c are positive real numbers less than 1, with $a+b+c=2$, then*

$$\left(\frac{a}{1-a}\right)\left(\frac{b}{1-b}\right)\left(\frac{c}{1-c}\right) \geq 8.$$

After performing the substitution $x = 1-a$, $y = 1-b$, $z = 1-c$, we obtain that $x+y+z = 3-(a+b+c) = 1$, $a = 1-x = y+z$, $b = z+x$, $c = x+y$. Hence the inequality is equivalent to

$$\left(\frac{y+z}{x}\right)\left(\frac{z+x}{y}\right)\left(\frac{x+y}{z}\right) \geq 8,$$

and in turn, this is equivalent to

$$(x+y)(y+z)(z+x) \geq 8xyz.$$

This last inequality is quite easy to prove. It is enough to apply three times the AM-GM inequality under the form $(x+y) \geq 2\sqrt{xy}$ (see Exercise 1.26).

It may be possible that the extra condition is used only as part of the solution, as in the following two examples.

Example 1.7.2. (Mexico, 2007) *If a, b, c are positive real numbers that satisfy $a+b+c=1$, prove that*

$$\sqrt{a+bc} + \sqrt{b+ca} + \sqrt{c+ab} \leq 2.$$

Using the condition $a+b+c=1$, we have that

$$a+bc=a(a+b+c)+bc=(a+b)(a+c),$$

then, by the AM-GM inequality it follows that

$$\sqrt{a+bc}=\sqrt{(a+b)(a+c)}\leq\frac{2a+b+c}{2}.$$

Similarly,

$$\sqrt{b+ca}\leq\frac{2b+c+a}{2}\qquad\text{and}\qquad\sqrt{c+ab}\leq\frac{2c+a+b}{2}.$$

Thus, after adding the three inequalities we obtain

$$\begin{aligned}&\sqrt{a+bc}+\sqrt{b+ca}+\sqrt{c+ab}\\&\leq\frac{2a+b+c}{2}+\frac{2b+c+a}{2}+\frac{2c+a+b}{2}=\frac{4a+4b+4c}{2}=2.\end{aligned}$$

The equality holds when $a+b=a+c$, $b+c=b+a$ and $c+a=c+b$, that is, when $a=b=c=\frac{1}{3}$.

Example 1.7.3. *If a, b, c are positive real numbers with $ab+bc+ca=1$, prove that*

$$\frac{a}{\sqrt{a^2+1}}+\frac{b}{\sqrt{b^2+1}}+\frac{c}{\sqrt{c^2+1}}\leq\frac{3}{2}.$$

Note that $(a^2+1)=a^2+ab+bc+ca=(a+b)(a+c)$. Similarly, $b^2+1=(b+c)(b+a)$ and $c^2+1=(c+a)(c+b)$. Now, the inequality under consideration is equivalent to

$$\frac{a}{\sqrt{(a+b)(a+c)}}+\frac{b}{\sqrt{(b+c)(b+a)}}+\frac{c}{\sqrt{(c+a)(c+b)}}\leq\frac{3}{2}.$$

Using the AM-GM inequality, applied to every element of the sum on the left-hand side, we obtain

$$\begin{aligned}&\frac{a}{\sqrt{(a+b)(a+c)}}+\frac{b}{\sqrt{(b+c)(b+a)}}+\frac{c}{\sqrt{(c+a)(c+b)}}\\&\leq\frac{1}{2}\left(\frac{a}{a+b}+\frac{a}{a+c}\right)+\frac{1}{2}\left(\frac{b}{b+c}+\frac{b}{b+a}\right)+\frac{1}{2}\left(\frac{c}{c+a}+\frac{c}{c+b}\right)=\frac{3}{2}.\end{aligned}$$

Many inequality problems suggest which substitution should be made. In the following example the substitution allows us to make at least one of the terms in the inequality look simpler.

Example 1.7.4. (India, 2002) *If a, b, c are positive real numbers, prove that*

$$\frac{a}{b}+\frac{b}{c}+\frac{c}{a}\geq\frac{c+a}{c+b}+\frac{a+b}{a+c}+\frac{b+c}{b+a}.$$

Making the substitution $x = \frac{a}{b}$, $y = \frac{b}{c}$, $z = \frac{c}{a}$ the left-hand side of the inequality is now more simple, $x + y + z$. Let us see how the right-hand side changes. The first element of the sum is modified as follows:

$$\frac{c+a}{c+b} = \frac{1+\frac{a}{c}}{1+\frac{b}{c}} = \frac{1+\frac{a}{b}\frac{b}{c}}{1+\frac{b}{c}} = \frac{1+xy}{1+y} = x + \frac{1-x}{1+y}.$$

Similarly,

$$\frac{a+b}{a+c} = y + \frac{1-y}{1+z} \quad \text{and} \quad \frac{b+c}{b+a} = z + \frac{1-z}{1+x}.$$

Now, the inequality is equivalent to

$$\frac{x-1}{1+y} + \frac{y-1}{1+z} + \frac{z-1}{1+x} \geq 0$$

with the extra condition $xyz = 1$.

The last inequality can be rewritten as

$$(x^2-1)(z+1) + (y^2-1)(x+1) + (z^2-1)(y+1) \geq 0,$$

which in turn is equivalent to

$$x^2z + y^2x + z^2y + x^2 + y^2 + z^2 \geq x + y + z + 3.$$

But, from the AM-GM inequality, we have $x^2z+y^2x+z^2y \geq 3\sqrt[3]{x^3y^3z^3} = 3$. Also, $x^2+y^2+z^2 \geq \frac{1}{3}(x+y+z)^2 = \frac{x+y+z}{3}(x+y+z) \geq \sqrt[3]{xyz}(x+y+z) = x+y+z$, where the first inequality follows from inequality (1.11).

In order to make a substitution, sometimes it is necessary to work a little bit beforehand, as we can see in the following example. This example also helps us to point out that we may need to make more than one substitution in the same problem.

Example 1.7.5. *Let a, b, c be positive real numbers, prove that*

$$(a+b)(a+c) \geq 2\sqrt{abc(a+b+c)}.$$

Dividing both sides of the given inequality by a^2 and setting $x = \frac{b}{a}$, $y = \frac{c}{a}$, the inequality becomes

$$(1+x)(1+y) \geq 2\sqrt{xy(1+x+y)}.$$

Now, dividing both sides by xy and making the substitution $r = 1+\frac{1}{x}$, $s = 1+\frac{1}{y}$, the inequality we need to prove becomes

$$rs \geq 2\sqrt{rs-1}.$$

This last inequality is equivalent to $(rs-2)^2 \geq 0$, which become evident after squaring both sides and doing some algebra.

It is a common situation for inequality problems to have several solutions and also to accept several substitutions that help to solve the problem. We will see an instance of this in the next example.

Example 1.7.6. (Korea, 1998) *If a, b, c are positive real numbers such that $a+b+c=abc$, prove that*

$$\frac{1}{\sqrt{1+a^2}}+\frac{1}{\sqrt{1+b^2}}+\frac{1}{\sqrt{1+c^2}}\leq\frac{3}{2}.$$

Under the substitution $x=\frac{1}{a}$, $y=\frac{1}{b}$, $z=\frac{1}{c}$, condition $a+b+c=abc$ becomes $xy+yz+zx=1$ and the inequality becomes equivalent to

$$\frac{x}{\sqrt{x^2+1}}+\frac{y}{\sqrt{y^2+1}}+\frac{z}{\sqrt{z^2+1}}\leq\frac{3}{2}.$$

This is the third example in this section.

Another solution is to make the substitution $a=\tan A$, $b=\tan B$, $c=\tan C$. Since $\tan A+\tan B+\tan C=\tan A\tan B\tan C$, then $A+B+C=\pi$ (or a multiple of π). Now, since $1+\tan^2 A=(\cos A)^{-2}$, the inequality is equivalent to $\cos A+\cos B+\cos C\leq\frac{3}{2}$, which is a valid result as will be shown in Example 2.5.2. Note that the Jensen inequality cannot be applied in this case because the function $f(x)=\frac{1}{\sqrt{1+x^2}}$ is not concave in $\mathbb{R}^+$.

We note that not all substitutions are algebraic, since there are trigonometric substitutions that can be useful, as is shown in the last example and as we will see next. Also, as will be shown in Sections 2.2 and 2.5 of the next chapter, there are some geometric substitutions that can be used for the same purposes.

Example 1.7.7. (Romania, 2002) *If a, b, c are real numbers in the interval $(0,1)$, prove that*

$$\sqrt{abc}+\sqrt{(1-a)(1-b)(1-c)}<1.$$

Making the substitution $a=\cos^2 A$, $b=\cos^2 B$, $c=\cos^2 C$, with A, B, C in the interval $(0,\frac{\pi}{2})$, we obtain that $\sqrt{1-a}=\sqrt{1-\cos^2 A}=\sin A$, $\sqrt{1-b}=\sin B$ and $\sqrt{1-c}=\sin C$. Therefore the inequality is equivalent to

$$\cos A\cos B\cos C+\sin A\sin B\sin C<1.$$

But observe that

$$\begin{aligned}\cos A\cos B\cos C+\sin A\sin B\sin C&<\cos A\cos B+\sin A\sin B\\&=\cos(A-B)\leq 1.\end{aligned}$$

Exercise 1.109. Let x, y, z be positive real numbers. Prove that

$$\frac{x^3}{x^3+2y^3}+\frac{y^3}{y^3+2z^3}+\frac{z^3}{z^3+2x^3}\geq 1.$$

Exercise 1.110. (Kazakhstan, 2008) Let x, y, z be positive real numbers such that $xyz=1$. Prove that

$$\frac{1}{yz+z}+\frac{1}{zx+x}+\frac{1}{xy+y}\geq\frac{3}{2}.$$

Exercise 1.111. (Russia, 2004) If $n > 3$ and $x_1, x_2, \ldots, x_n$ are positive real numbers with $x_1 x_2 \cdots x_n = 1$, prove that

$$\frac{1}{1+x_1+x_1x_2} + \frac{1}{1+x_2+x_2x_3} + \cdots + \frac{1}{1+x_n+x_nx_1} > 1.$$

Exercise 1.112. (Poland, 2006) Let a, b, c be positive real numbers such that $ab + bc + ca = abc$. Prove that

$$\frac{a^4+b^4}{ab(a^3+b^3)} + \frac{b^4+c^4}{bc(b^3+c^3)} + \frac{c^4+a^4}{ca(c^3+a^3)} \geq 1.$$

Exercise 1.113. (Ireland, 2007) Let a, b, c be positive real numbers, prove that

$$\frac{1}{3}\left(\frac{bc}{a} + \frac{ca}{b} + \frac{ca}{b}\right) \geq \sqrt{\frac{a^2+b^2+c^2}{3}} \geq \frac{a+b+c}{3}.$$

Exercise 1.114. (Romania, 2008) Let a, b, c be positive real numbers with $abc = 8$. Prove that

$$\frac{a-2}{a+1} + \frac{b-2}{b+1} + \frac{c-2}{c+1} \leq 0.$$

1.8 Muirhead's theorem

In 1903, R.F. Muirhead published a paper containing the study of some algebraic methods applicable to identities and inequalities of symmetric algebraic functions of n variables.

While considering algebraic expressions of the form $x_1^{a_1} x_2^{a_2} \cdots x_n^{a_n}$, he analyzed symmetric polynomials containing these expressions in order to create a "certain order" in the space of n-tuples $(a_1, a_2, \ldots, a_n)$ satisfying the condition $a_1 \geq a_2 \geq \cdots \geq a_n$.

We will assume that $x_i > 0$ for all $1 \leq i \leq n$. We will denote by

$$\sum\nolimits_! F(x_1, \ldots, x_n)$$

the sum of the $n!$ terms obtained from evaluating F in all possible permutations of $(x_1, \ldots, x_n)$. We will consider only the particular case

$$F(x_1, \ldots, x_n) = x_1^{a_1} x_2^{a_2} \cdots x_n^{a_n} \quad \text{with } x_i > 0,\ a_i \geq 0.$$

We write $[a] = [a_1, a_2, \ldots, a_n] = \frac{1}{n!}\sum_! x_1^{a_1} x_2^{a_2} \cdots x_n^{a_n}$. For instance, for the variables x, y, $z > 0$ we have that

$$[1,1] = xy, \;\; [1,1,1] = xyz, \;\; [2,1,0] = \frac{1}{3!}[x^2(y+z) + y^2(x+z) + z^2(x+y)].$$

It is clear that $[a]$ is invariant under any permutation of the $(a_1, a_2, \ldots, a_n)$ and therefore two sets of a are the same if they only differ in arrangement. We will say that a mean value of the type $[a]$ is a symmetrical mean. In particular, $[1, 0, \ldots, 0] = \frac{(n-1)!}{n!}(x_1 + x_2 + \cdots + x_n) = \frac{1}{n}\sum_{i=1}^{n} x_i$ is the arithmetic mean and $[\frac{1}{n}, \frac{1}{n}, \ldots, \frac{1}{n}] = \frac{n!}{n!}(x_1^{\frac{1}{n}} \cdot x_2^{\frac{1}{n}} \cdots x_n^{\frac{1}{n}}) = \sqrt[n]{x_1 x_2 \cdots x_n}$ is the geometric mean. When $a_1 + a_2 + \cdots + a_n = 1$, $[a]$ is a common generalization of both the arithmetic mean and the geometric mean.

If $a_1 \geq a_2 \geq \cdots \geq a_n$ and $b_1 \geq b_2 \geq \cdots \geq b_n$, usually $[b]$ is not comparable to $[a]$, in the sense that there is an inequality between their associated expressions valid for all n-tuples of non-negative real numbers $x_1, x_2, \ldots, x_n$.

Muirhead wanted to compare the values of the symmetric polynomials $[a]$ and $[b]$ for any set of non-negative values of the variables occurring in both polynomials.

From now on we denote $(a) = (a_1, a_2, \ldots, a_n)$.

Definition 1.8.1. *We will say that $(b) \prec (a)$ ((b) is majorized by (a)) when (a) and (b) can be rearranged to satisfy the following two conditions:*

(1) $\displaystyle\sum_{i=1}^{n} b_i = \sum_{i=1}^{n} a_i$;

(2) $\displaystyle\sum_{i=1}^{\nu} b_i \leq \sum_{i=1}^{\nu} a_i$ *for all* $1 \leq \nu < n$.

It is clear that $(a) \prec (a)$ and that $(b) \prec (a)$ and $(c) \prec (b)$ implies $(c) \prec (a)$.

Theorem 1.8.2 (Muirhead's theorem). $[b] \leq [a]$ *for any n-tuple of non-negative numbers $(x_1, x_2, \ldots, x_n)$ if and only if $(b) \prec (a)$. Equality takes place only when (b) and (a) are identical or when all the x_is are equal.*

Before going through the proof, which is quite difficult, let us look at some examples. First, it is clear that $[2, 0, 0]$ cannot be compared with $[1, 1, 1]$ because the first condition in Definition 1.8.1 is not satisfied, but we can see that $[2, 0, 0] \geq [1, 1, 0]$, which is equivalent to

$$x^2 + y^2 + z^2 \geq xy + yz + zx.$$

In the same way, we can see that

1. $x^2 + y^2 \geq 2xy \Leftrightarrow [2, 0] \geq [1, 1]$,
2. $x^3 + y^3 + z^3 \geq 3xyz \Leftrightarrow [3, 0, 0] \geq [1, 1, 1]$,
3. $x^5 + y^5 \geq x^3y^2 + x^2y^3 \Leftrightarrow [5, 0] \geq [3, 2]$,
4. $x^2y^2 + y^2z^2 + z^2x^2 \geq x^2yz + y^2xz + z^2xy \Leftrightarrow [2, 2, 0] \geq [2, 1, 1]$,

and all these inequalities are satisfied if we take for granted Muirhead's theorem.

Proof of Muirhead's theorem. Suppose that $[b] \leq [a]$ for any n positive numbers $x_1, x_2, \ldots, x_n$. Taking $x_i = x$, for all i, we obtain

$$x^{\sum b_i} = [b] \leq [a] = x^{\sum a_i}.$$

This can only be true for all x if $\sum b_i = \sum a_i$.
Next, take $x_1 = x_2 = \cdots = x_\nu = x$, $x_{\nu+1} = \cdots = x_n = 1$ and x very large. Since (b) and (a) are in decreasing order, the index of the highest powers of x in $[b]$ and $[a]$ are

$$b_1 + b_2 + \cdots + b_\nu, \quad a_1 + a_2 + \cdots + a_\nu,$$

respectively. Thus, it is clear that the first sum can not be greater than the second and this proves (2) in Definition 1.8.1.

The proof in the other direction is more difficult to establish, and we will need a new definition and two more lemmas.

We define a special type of linear transformation T of the a's, as follows. Suppose that $a_k > a_l$, then let us write

$$a_k = \rho + \tau, \quad a_l = \rho - \tau \quad (0 < \tau \leq \rho).$$

If now $0 \leq \sigma < \tau \leq \rho$, then a T-transformation is defined by

$$T(a_k) = b_k = \rho + \sigma = \frac{\tau + \sigma}{2\tau} a_k + \frac{\tau - \sigma}{2\tau} a_l,$$

$$T(a_l) = b_l = \rho - \sigma = \frac{\tau - \sigma}{2\tau} a_k + \frac{\tau + \sigma}{2\tau} a_l,$$

$$T(a_\nu) = a_\nu \quad (\nu \neq k,\, \nu \neq l).$$

If (b) arises from (a) by a T-transformation, we write $b = Ta$. The definition does not necessarily imply that either the (a) or the (b) are in decreasing order.
The sufficiency of our comparability condition will be established if we can prove the following two lemmas.

Lemma 1.8.3. *If $b = Ta$, then $[b] \leq [a]$ with equality taking place only when all the x_i's are equal.*

Proof. We may rearrange (a) and (b) so that $k = 1$, $l = 2$. Thus

$$\begin{aligned}
[a] - [b] &= [\rho + \tau, \rho - \tau, a_3, \ldots] - [\rho + \sigma, \rho - \sigma, a_3, \ldots] \\
&= \frac{1}{2n!} \sum\nolimits_! x_3^{a_3} \cdots x_n^{a_n} (x_1^{\rho+\tau} x_2^{\rho-\tau} + x_1^{\rho-\tau} x_2^{\rho+\tau}) \\
&\quad - \frac{1}{2n!} \sum\nolimits_! x_3^{a_3} \cdots x_n^{a_n} (x_1^{\rho+\sigma} x_2^{\rho-\sigma} + x_1^{\rho-\sigma} x_2^{\rho+\sigma}) \\
&= \frac{1}{2n!} \sum\nolimits_! (x_1 x_2)^{\rho-\tau} x_3^{a_3} \cdots x_n^{a_n} (x_1^{\tau+\sigma} - x_2^{\tau+\sigma})(x_1^{\tau-\sigma} - x_2^{\tau-\sigma}) \geq 0
\end{aligned}$$

with equality being the case only when all the x_i's are equal. □

Lemma 1.8.4. *If $(b) \prec (a)$, but (b) is not identical to (a), then (b) can be derived from (a) using the successive application of a finite number of T-transformations.*

Proof. We call the number of differences $a_\nu - b_\nu$ which are not zero, the *discrepancy* between (a) and (b). If the discrepancy is zero, the sets are identical. We will prove the lemma by induction, assuming it to be true when the discrepancy is less than r and proving that it is also true when the discrepancy is r.

Suppose then that $(b) \prec (a)$ and that the discrepancy is $r > 0$. Since $\sum_{i=1}^n a_i = \sum_{i=1}^n b_i$, and $\sum(a_\nu - b_\nu) = 0$, and not all of these differences are zero, there must be positive and negative differences, and the first which is not zero must be positive because of the second condition of $(b) \prec (a)$. We can therefore find k and l such that

$$b_k < a_k, \quad b_{k+1} = a_{k+1}, \quad \dots \quad , b_{l-1} = a_{l-1}, \quad b_l > a_l; \tag{1.12}$$

that is, $a_l - b_l$ is the first negative difference and $a_k - b_k$ is the last positive difference which precedes it.

We take $a_k = \rho + \tau$, $a_l = \rho - \tau$, and define σ by

$$\sigma = \max(|b_k - \rho| \ , \ |b_l - \rho|).$$

Then $0 < \tau \leq \rho$, since $a_k > a_l$. Also, one (possible both) of $b_l - \rho = -\sigma$ or $b_k - \rho = \sigma$ is true, since $b_k \geq b_l$, and $\sigma < \tau$, since $b_k < a_k$ and $b_l > a_l$. Hence $0 \leq \sigma < \tau \leq \rho$.

We now write $a'_k = \rho + \sigma$, $a'_l = \rho - \sigma$, $a'_\nu = a_\nu$ $(\nu \neq k,\ \nu \neq l)$. If $b_k - \rho = \sigma$, $a'_k = b_k$, and if $b_l - \rho = -\sigma$, then $a'_l = b_l$. Since the pairs a_k, b_k and a_l, b_l each contributes one unit to the discrepancy r between (b) and (a), the discrepancy between (b) and (a') is smaller, being equal to $r - 1$ or $r - 2$.

Next, comparing the definition of (a') with the definition of the T-transformation, and observing that $0 \leq \sigma < \tau \leq \rho$, we can infer that (a') arises from (a) by a T-transformation.

Finally, let us prove that $(b) \prec (a')$. In order to do that, we must verify that the two conditions of $\prec$ are satisfied and that the order of (a') is non-increasing. For the first one, we have

$$a'_k + a'_l = 2\rho = a_k + a_l, \quad \sum_{i=1}^n b_i = \sum_{i=1}^n a_i = \sum_{i=1}^n a'_i.$$

For the second one, we must prove that

$$b_1 + b_2 + \cdots + b_\nu \leq a'_1 + a'_2 + \cdots + a'_\nu.$$

Now, this is true if $\nu < k$ or $\nu \geq l$, as can be established by using the definition of (a') and also the second condition of $(b) \prec (a)$. It is true for $\nu = k$, because it is true for $\nu = k - 1$ and $b_k \leq a'_k$, and it is true for $k < \nu < l$ because it is valid for $\nu = k$ and the intervening b and a' are identical.

Finally, we observe that

$$b_k \le \rho + |b_k - \rho| \le \rho + \sigma = a'_k,$$

$$b_l \ge \rho - |b_l - \rho| \ge \rho - \sigma = a'_l,$$

and then, using (1.12),

$$a'_{k-1} = a_{k-1} \ge a_k = \rho + \tau > \rho + \sigma = a'_k \ge b_k \ge b_{k+1} = a_{k+1} = a'_{k+1},$$

$$a'_{l-1} = a_{l-1} = b_{l-1} \ge b_l \ge a'_l = \rho - \sigma > \rho - \tau = a_l \ge a_{l+1} = a'_{l+1}.$$

The inequalities involving a' are as required.

We have thus proved that $(b) \prec (a')$, a set arising from (a) using a transformation T and having a discrepancy from (b) of less than r. This proves the lemma and completes the proof of Muirhead's theorem. □

The proof of Muirhead's theorem demonstrates to us how the difference between two comparable means can be decomposed as a sum of obviously positive terms by repeated application of the T-transformation. We can produce from this result a new proof for the AM-GM inequality.

Example 1.8.5 (The AM-GM inequality). *For real positive numbers $y_1, y_2, \ldots, y_n$,*

$$\frac{y_1 + y_2 + \cdots + y_n}{n} \ge \sqrt[n]{y_1 y_2 \cdots y_n}.$$

Note that the AM-GM inequality is equivalent to

$$\frac{1}{n}\sum_{i=1}^{n} x_i^n \ge x_1 x_2 \cdots x_n,$$

where $x_i = \sqrt[n]{y_i}$.

Now, we observe that

$$\frac{1}{n}\sum_{i=1}^{n} x_i^n = [n, 0, 0, \ldots, 0] \quad \text{and} \quad x_1 x_2 \cdots x_n = [1, 1, \ldots, 1].$$

By Muirhead's theorem we can show that

$$[n, 0, 0, \ldots, 0] \ge [1, 1, \ldots, 1].$$

Next, we provide another proof for the AM-GM inequality, something we shall do by following the ideas inherent in the proof of Muirhead's theorem in order to illustrate how it works.

$$\begin{aligned}\frac{1}{n}\sum_{i=1}^{n} x_i^n - (x_1x_2\cdots x_n) &= [n,0,0,\ldots,0] - [1,1,\ldots,1]\\
&= ([n,0,0,\ldots,0] - [n-1,1,0,\ldots,0])\\
&\quad + ([n-1,1,0,\ldots,0] - [n-2,1,1,0,\ldots,0])\\
&\quad + ([n-2,1,1,0,\ldots,0] - [n-3,1,1,1,0,\ldots,0])\\
&\quad + \cdots + ([2,1,1,\ldots,1] - [1,1,\ldots 1])\\
&= \frac{1}{2n!}\Big(\sum\nolimits_! (x_1^{n-1} - x_2^{n-1})(x_1 - x_2)\\
&\quad + \sum\nolimits_! (x_1^{n-2} - x_2^{n-2})(x_1 - x_2)x_3\\
&\quad + \sum\nolimits_! (x_1^{n-3} - x_2^{n-3})(x_1 - x_2)x_3x_4 + \cdots\Big).\end{aligned}$$

Since $(x_r^\nu - x_s^\nu)(x_r - x_s) > 0$, unless $x_r = x_s$, the inequality follows.

Example 1.8.6. *If a, b are positive real numbers, then*

$$\sqrt{\frac{a^2}{b}} + \sqrt{\frac{b^2}{a}} \geq \sqrt{a} + \sqrt{b}.$$

Setting $x = \sqrt{a}$, $y = \sqrt{b}$ and simplifying, we have to prove

$$x^3 + y^3 \geq xy(x+y).$$

Using Muirhead's theorem, we get

$$[3,0] = \frac{1}{2}(x^3 + y^3) \geq \frac{1}{2}xy(x+y) = [2,1],$$

and thus the result follows.

Example 1.8.7. *If a, b, c are non-negative real numbers, prove that*

$$a^3 + b^3 + c^3 + abc \geq \frac{1}{7}(a+b+c)^3.$$

It is not difficult to see that

$$(a+b+c)^3 = 3[3,0,0] + 18[2,1,0] + 36[1,1,1].$$

Then we need to prove that

$$3[3,0,0] + 6[1,1,1] \geq \frac{1}{7}(3[3,0,0] + 18[2,1,0] + 36[1,1,1]),$$

that is,

$$\frac{18}{7}[3,0,0] + \left(6 - \frac{36}{7}\right)[1,1,1] \geq \frac{18}{7}[2,1,0]$$

or

$$\frac{18}{7}([3,0,0]-[2,1,0])+\left(6-\frac{36}{7}\right)[1,1,1]\geq 0.$$

This follows using the inequalities $[3,0,0]\geq[2,1,0]$ and $[1,1,1]\geq 0$.

Example 1.8.8. *If a, b, c are non-negative real numbers, prove that*

$$a+b+c\leq\frac{a^2+b^2}{2c}+\frac{b^2+c^2}{2a}+\frac{c^2+a^2}{2b}\leq\frac{a^3}{bc}+\frac{b^3}{ca}+\frac{c^3}{ab}.$$

The inequalities are equivalent to the following:

$$2(a^2bc+ab^2c+abc^2)\leq ab(a^2+b^2)+bc(b^2+c^2)+ca(c^2+a^2)\leq 2(a^4+b^4+c^4),$$

which is in turn equivalent to $[2,1,1]\leq[3,1,0]\leq[4,0,0]$. Using Muirhead's theorem we arrive at the result.

Exercise 1.115. Prove that any three positive real numbers a, b and c satisfy

$$a^5+b^5+c^5\geq a^3bc+b^3ca+c^3ab.$$

Exercise 1.116. (IMO, 1961) Let a, b, c be the lengths of the sides of a triangle, and let (ABC) denote its area. Prove that

$$4\sqrt{3}(ABC)\leq a^2+b^2+c^2.$$

Exercise 1.117. Let a, b, c be positive real numbers. Prove that

$$\frac{a}{(a+b)(a+c)}+\frac{b}{(b+c)(b+a)}+\frac{c}{(c+a)(c+b)}\leq\frac{9}{4(a+b+c)}.$$

Exercise 1.118. (IMO, 1964) Let a, b, c be positive real numbers. Prove that

$$a^3+b^3+c^3+3abc\geq ab(a+b)+bc(b+c)+ca(c+a).$$

Exercise 1.119. (Short list Iberoamerican, 2003) Let a, b, c be positive real numbers. Prove that

$$\frac{a^3}{b^2-bc+c^2}+\frac{b^3}{c^2-ca+a^2}+\frac{c^3}{a^2-ab+b^2}\geq a+b+c.$$

Exercise 1.120. (Short list IMO, 1998) Let a, b, c be positive real numbers such that $abc=1$. Prove that

$$\frac{a^3}{(1+b)(1+c)}+\frac{b^3}{(1+c)(1+a)}+\frac{c^3}{(1+a)(1+b)}\geq\frac{3}{4}.$$

Chapter 2

Geometric Inequalities

2.1 Two basic inequalities

The two basic geometric inequalities we will be refering to in this section involve triangles. One of them is the triangle inequality and we will refer to it as D1; the second one is not really an inequality, but it represents an important observation concerning the geometry of triangles which points out that if we know the greatest angle of a triangle, then we know which is the longest side of the triangle; this observation will be denoted as D2.

D1. *If A, B and C are points on the plane, then*

$$AB + BC \geq AC.$$

Moreover, the equality holds if and only if B lies on the line segment AC.

D2. *In a triangle, the longest side is opposite to the greatest angle and vice versa.*

Hence, if in the triangle ABC we have $\angle A > \angle B$, then $BC > CA$.

Exercise 2.1. (i) If a, b, c are positive numbers with $a < b + c$, $b < c + a$ and $c < a + b$, then a triangle exists with side lengths a, b and c.

(ii) To be able to construct a triangle with side lengths $a \leq b \leq c$, it is sufficient that $c < a + b$.

(iii) It is possible to construct a triangle with sides of length a, b and c if and only if there are positive numbers x, y, z such that $a = x + y$, $b = y + z$ and $c = z + x$.

Exercise 2.2. (i) If it is possible to construct a triangle with side-lengths $a < b < c$, then it is possible to construct a triangle with side-lengths $\sqrt{a} < \sqrt{b} < \sqrt{c}$.

(ii) The converse of (i) is false.

(iii) If it is possible to construct a triangle with side-lengths $a < b < c$, then it is possible to construct a triangle with side-lengths $\frac{1}{a+b}$, $\frac{1}{b+c}$ and $\frac{1}{c+a}$.

Exercise 2.3. Let a, b, c, d and e be the lengths of five segments such that it is possible to construct a triangle using any three of them. Prove that there are three of them that form an acute triangle.

Sometimes the key to solve a problem lies in the ability to identify certain quantities that can be related to geometric measurements, as in the following example.

Example 2.1.1. *If a, b, c are positive numbers with $a^2 + b^2 - ab = c^2$, prove that $(a - b)(b - c) \leq 0$.*

Since $c^2 = a^2 + b^2 - ab = a^2 + b^2 - 2ab \cos 60°$, we can think that a, b, c are the lengths of the sides of a triangle such that the measure of the angle opposed to the side of length c is $60°$. The angles of the triangle ABC satisfy $\angle A \leq 60°$ and $\angle B \geq 60°$, or $\angle A \geq 60°$ and $\angle B \leq 60°$; hence, using property D2 we can deduce that $a \leq c \leq b$ or $a \geq c \geq b$. In any case it follows that $(a - b)(b - c) \leq 0$.

Observation 2.1.2. *We can also solve the example above without the identification of a, b and c with the lengths of the sides of a triangle.*

First suppose that $a \leq b$, then the fact that $a^2 + b^2 - ab = c^2$ implies that $a(a-b) = c^2 - b^2 = (c-b)(c+b)$, hence $c - b \leq 0$ and therefore $(a-b)(b-c) \leq 0$.

Similarly, $a \geq b$ implies $c - b \geq 0$, and hence

$$(a - b)(b - c) \leq 0.$$

Another situation where it is not obvious that we can identify the elements with a geometric inequality, or that the use of geometry may be helpful, is shown in the following example.

Example 2.1.3. *If a, b, c are positive numbers, then*

$$\sqrt{a^2 + ac + c^2} \leq \sqrt{a^2 - ab + b^2} + \sqrt{b^2 - bc + c^2}.$$

The radicals suggest using the cosine law with angles of $120°$ and of $60°$ as follows: $a^2 + ac + c^2 = a^2 + c^2 - 2ac \cos 120°$, $a^2 - ab + b^2 = a^2 + b^2 - 2ab \cos 60°$ and $b^2 - bc + c^2 = b^2 + c^2 - 2bc \cos 60°$.

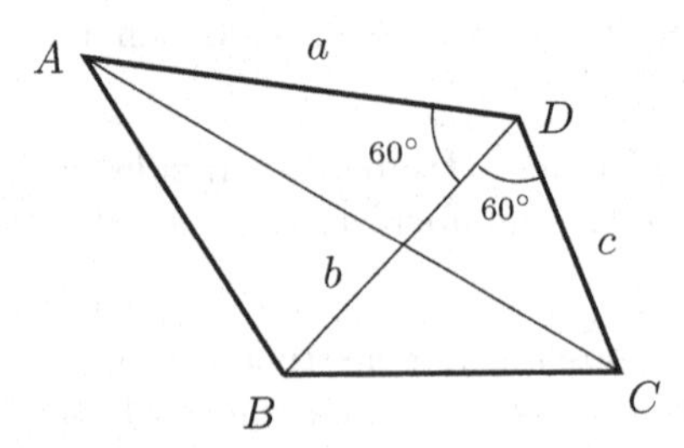

Then, if we consider a quadrilateral $ABCD$, with $\angle ADB = \angle BDC = 60°$ and $\angle ADC = 120°$, such that $AD = a$, $BD = b$ and $CD = c$, we can deduce that $AB = \sqrt{a^2 - ab + b^2}$, $BC = \sqrt{b^2 - bc + c^2}$ and $CA = \sqrt{a^2 + ac + c^2}$. The inequality we have to prove becomes the triangle inequality for the triangle ABC.

Exercise 2.4. Let ABC be a triangle with $\angle A > \angle B$, prove that $BC > \frac{1}{2}AB$.

Exercise 2.5. Let $ABCD$ be a convex quadrilateral, prove that

(i) if $AB + BD < AC + CD$, then $AB < AC$,

(ii) if $\angle A > \angle C$ and $\angle D > \angle B$, then $BC > \frac{1}{2}AD$.

Exercise 2.6. If a_1, a_2, a_3, a_4 and a_5 are the lengths of the sides of a convex pentagon and if d_1, d_2, d_3, d_4 and d_5 are the lengths of its diagonals, prove that

$$\frac{1}{2} < \frac{a_1 + a_2 + a_3 + a_4 + a_5}{d_1 + d_2 + d_3 + d_4 + d_5} < 1.$$

Exercise 2.7. The length m_a of the median AA' of a triangle ABC satisfies $m_a > \frac{b+c-a}{2}$.

Exercise 2.8. If the length m_a of the median AA' of a triangle ABC satisfies $m_a > \frac{1}{2}a$, prove that $\angle BAC < 90°$.

Exercise 2.9. If AA' is the median of the triangle ABC and if $AB < AC$, then $\angle BAA' > \angle A'AC$.

Exercise 2.10. If m_a, m_b and m_c are the lengths of the medians of a triangle with side-lengths a, b and c, respectively, prove that it is possible to construct a triangle with side-lengths m_a, m_b and m_c, and that

$$\frac{3}{4}(a + b + c) < m_a + m_b + m_c < a + b + c.$$

Exercise 2.11. (Ptolemy's inequality) If $ABCD$ is a convex quadrilateral, then $AC \cdot BD \leq AB \cdot CD + BC \cdot DA$. The equality holds if and only if $ABCD$ is a cyclic quadrilateral.

Exercise 2.12. Let $ABCD$ be a cyclic quadrilateral. Prove that $AC > BD$ if and only if $(AD - BC)(AB - DC) > 0$.

Exercise 2.13. (Pompeiu's problem) Let ABC be an equilateral triangle and let P be a point that does not belong to the circumcircle of ABC. Prove that PA, PB and PC are the lengths of the sides of a triangle.

Exercise 2.14. If $ABCD$ is a paralelogram, prove that

$$|AB^2 - BC^2| < AC \cdot BD.$$

Exercise 2.15. If a, b and c are the lengths of the sides of a triangle, m_a, m_b and m_c represent the lengths of the medians and R is the circumradius, prove that

(i) $\dfrac{a^2+b^2}{m_c}+\dfrac{b^2+c^2}{m_a}+\dfrac{c^2+a^2}{m_b}\leq 12R$,

(ii) $m_a(bc-a^2)+m_b(ca-b^2)+m_c(ab-c^2)\geq 0$.

Exercise 2.16. Let ABC be a triangle whose sides have lengths a, b and c. Suppose that $c>b$, prove that

$$\frac{1}{2}(c-b)<m_b-m_c<\frac{3}{2}(c-b),$$

where m_b and m_c are the lengths of the medians.

Exercise 2.17. (Iran, 2005) Let ABC be a triangle with $\angle A=90°$. Let D be the intersection of the internal angle bisector of $\angle A$ with the side BC and let I_a be the center of the excircle of the triangle ABC opposite to the vertex A. Prove that

$$\frac{AD}{DI_a}\leq\sqrt{2}-1.$$

2.2 Inequalities between the sides of a triangle

Inequalities involving the lengths of the sides of a triangle appear frequently in mathematical competitions. One sort of problems consists of those where you are asked to prove some inequality that is satisfied by the lengths of the sides of a triangle without any other geometric elements being involved, as in the following example.

Example 2.2.1. *The lengths a, b and c of the sides of a triangle satisfy*

$$a\,(b+c-a)<2bc.$$

Since the inequality is symmetric in b and c, we can assume, without loss of generality, that $c\leq b$. We will prove the inequality in the following cases.

Case 1. $a\leq b$.

Since they are the lengths of the sides of a triangle, we have that $b<a+c$; then

$$b+c-a=b-a+c<c+c=2c\leq\frac{2bc}{a}.$$

Case 2. $a\geq b$.

In this case $b-a\leq 0$, and since $a<b+c\leq 2b$, we can deduce that

$$b+c-a=c+b-a\leq c<\frac{2bc}{a}.$$

Another type of problem involving the lengths of the sides of a triangle is when we are asked to prove that a certain relationship between the numbers a, b and c is sufficient to construct a triangle with sides of the same length.

Example 2.2.2. (i) *If a, b, c are positive numbers and satisfy, $\left(a^2+b^2+c^2\right)^2 > 2\left(a^4+b^4+c^4\right)$, then a, b and c are the lengths of the sides of a triangle.*

(ii) *If a, b, c, d are positive numbers and satisfy*

$$\left(a^2+b^2+c^2+d^2\right)^2 > 3\left(a^4+b^4+c^4+d^4\right),$$

then,using any three of them we can construct a triangle.

For part (i), it is sufficient to observe that

$$\left(a^2+b^2+c^2\right)^2 - 2\left(a^4+b^4+c^4\right) = (a+b+c)(a+b-c)(a-b+c)(-a+b+c) > 0,$$

and then note that none of these factors is negative. Compare this with Example 1.2.5.

For part (ii), we can deduce that

$$\begin{aligned} 3\left(a^4+b^4+c^4+d^4\right) &< \left(a^2+b^2+c^2+d^2\right)^2 \\ &= \left(\frac{a^2+b^2+c^2}{2}+\frac{a^2+b^2+c^2}{2}+d^2\right)^2 \\ &\le \left\{\left(\frac{a^2+b^2+c^2}{2}\right)^2+\left(\frac{a^2+b^2+c^2}{2}\right)^2+d^4\right\}\left(\sqrt{3}\right)^2. \end{aligned}$$

The second inequality follows from the Cauchy-Schwarz inequality; hence, $a^4+b^4+c^4 < 2\frac{\left(a^2+b^2+c^2\right)^2}{4}$. Using the first part we can deduce that a, b and c can be used to construct a triangle. Since the argument we used is symmetric in a, b, c and d, we obtain the result.

There is a technique that helps to transform one inequality between the lengths of the sides of a triangle into an inequality between positive numbers (of course related to the sides).This is called the *Ravi transformation*.

If the incircle (I,r) of the triangle ABC is tangent to the sides BC, CA and AB at the points X, Y and Z, respectively, we have that $x = AZ = YA$, $y = ZB = BX$ and $z = XC = CY$.

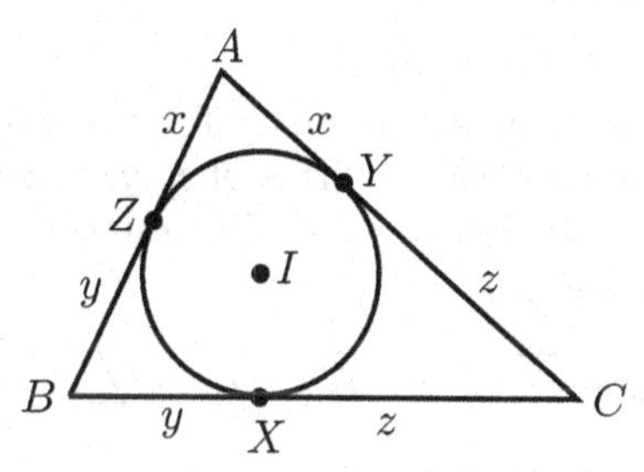

It is easily seen that $a = y+z$, $b = z+x$, $c = x+y$, $x = s-a$, $y = s-b$ and $z = s-c$, where $s = \frac{a+b+c}{2}$.

Let us now see how to use the Ravi transformation in the following example.

Example 2.2.3. *The lengths of the sides a, b and c of a triangle satisfy*

$$(b+c-a)(c+a-b)(a+b-c) \leq abc.$$

First, we have that

$$(b+c-a)(c+a-b)(a+b-c) = 8(s-a)(s-b)(s-c) = 8xyz,$$

on the other hand

$$abc = (x+y)(y+z)(z+x).$$

Thus, the inequality is equivalent to

$$8xyz \leq (x+y)(y+z)(z+x). \tag{2.1}$$

The last inequality follows from Exercise 1.26.

Example 2.2.4. (APMO, 1996) *Let a, b, c be the lengths of the sides of a triangle, prove that* $\sqrt{a+b-c} + \sqrt{b+c-a} + \sqrt{c+a-b} \leq \sqrt{a} + \sqrt{b} + \sqrt{c}$.

If we set $a = y+z$, $b = z+x$, $c = x+y$, we can deduce that $a+b-c = 2z$, $b+c-a = 2x$, $c+a-b = 2y$. Hence, the inequality is equivalent to

$$\sqrt{2x} + \sqrt{2y} + \sqrt{2z} \leq \sqrt{x+y} + \sqrt{y+z} + \sqrt{z+x}.$$

Now applying the inequality between the arithmetic mean and the quadratic mean (see Exercise 1.68), we get

$$\begin{aligned}
\sqrt{2x} + \sqrt{2y} + \sqrt{2z} &= \frac{\sqrt{2x}+\sqrt{2y}}{2} + \frac{\sqrt{2y}+\sqrt{2z}}{2} + \frac{\sqrt{2z}+\sqrt{2x}}{2} \\
&\leq \sqrt{\frac{2x+2y}{2}} + \sqrt{\frac{2y+2z}{2}} + \sqrt{\frac{2z+2x}{2}} \\
&= \sqrt{x+y} + \sqrt{y+z} + \sqrt{z+x}.
\end{aligned}$$

Moreover, the equality holds if and only if $x = y = z$, that is, if and only if $a = b = c$.

Also, it is possible to express the area of a triangle ABC, its inradius, its circumradius and its semiperimeter in terms of x, y, z. Since $a = x+y$, $b = y+z$ and $c = z+x$, we first obtain that $s = \frac{a+b+c}{2} = x+y+z$. Using Heron's formula for the area of a triangle, we get

$$(ABC) = \sqrt{s(s-a)(s-b)(s-c)} = \sqrt{(x+y+z)xyz}. \tag{2.2}$$

The formula $(ABC) = sr$ leads us to

$$r = \frac{(ABC)}{s} = \frac{\sqrt{(x+y+z)xyz}}{x+y+z} = \sqrt{\frac{xyz}{x+y+z}}.$$

Finally, from $(ABC) = \frac{abc}{4R}$ we get

$$R = \frac{(x+y)(y+z)(z+x)}{4\sqrt{(x+y+z)xyz}}.$$

Example 2.2.5. (India, 2003) *Let a, b, c be the side lengths of a triangle ABC. If we construct a triangle $A'B'C'$ with side lengths $a + \frac{b}{2}$, $b + \frac{c}{2}$, $c + \frac{a}{2}$, prove that $(A'B'C') \geq \frac{9}{4}(ABC)$.*

Since $a = y + z$, $b = z + x$ and $c = x + y$, the side lengths of the triangle $A'B'C'$ are $a' = \frac{x+2y+3z}{2}$, $b' = \frac{3x+y+2z}{2}$, $c' = \frac{2x+3y+z}{2}$. Using Heron's formula for the area of a triangle, we get

$$(A'B'C') = \sqrt{\frac{3(x+y+z)(2x+y)(2y+z)(2z+x)}{16}}.$$

Applying the *AM-GM* inequality to show that $2x+y \geq 3\sqrt[3]{x^2y}$, $2y+z \geq 3\sqrt[3]{y^2z}$, $2z+x \geq 3\sqrt[3]{z^2x}$, will help to reach the inequality

$$(A'B'C') \geq \sqrt{\frac{3(x+y+z)27(xyz)}{16}} = \frac{9}{4}(ABC).$$

Equation (2.2) establishes the last equality.

Exercise 2.18. Let a, b and c be the lengths of the sides of a triangle, prove that

$$3(ab+bc+ca) \leq (a+b+c)^2 \leq 4(ab+bc+ca).$$

Exercise 2.19. Let a, b and c be the lengths of the sides of a triangle, prove that

$$ab+bc+ca \leq a^2+b^2+c^2 \leq 2(ab+bc+ca).$$

Exercise 2.20. Let a, b and c be the lengths of the sides of a triangle, prove that

$$2\left(a^2+b^2+c^2\right) \leq (a+b+c)^2.$$

Exercise 2.21. Let a, b and c be the lengths of the sides of a triangle, prove that

$$\frac{3}{2} \leq \frac{a}{b+c} + \frac{b}{c+a} + \frac{c}{a+b} < 2.$$

Exercise 2.22. (IMO, 1964) Let a, b and c be the lengths of the sides of a triangle, prove that

$$a^2(b+c-a) + b^2(c+a-b) + c^2(a+b-c) \leq 3abc.$$

Exercise 2.23. Let a, b and c be the lengths of the sides of a triangle, prove that

$$a\left(b^2+c^2-a^2\right)+b\left(c^2+a^2-b^2\right)+c\left(a^2+b^2-c^2\right)\leq 3abc.$$

Exercise 2.24. (IMO, 1983) Let a, b and c be the lengths of the sides of a triangle, prove that

$$a^2b(a-b)+b^2c(b-c)+c^2a(c-a)\geq 0.$$

Exercise 2.25. Let a, b and c be the lengths of the sides of a triangle, prove that

$$\left|\frac{a-b}{a+b}+\frac{b-c}{b+c}+\frac{c-a}{c+a}\right|<\frac{1}{8}.$$

Exercise 2.26. The lengths a, b and c of the sides of a triangle satisfy $ab+bc+ca=3$. Prove that

$$3\leq a+b+c\leq 2\sqrt{3}.$$

Exercise 2.27. Let a, b, c be the lengths of the sides of a triangle, and let r be the inradius of the triangle. Prove that

$$\frac{1}{a}+\frac{1}{b}+\frac{1}{c}\leq\frac{\sqrt{3}}{2r}.$$

Exercise 2.28. Let a, b, c be the lengths of the sides of a triangle, and let s be the semiperimeter of the triangle. Prove that

(i) $(s-a)(s-b)<ab$,

(ii) $(s-a)(s-b)+(s-b)(s-c)+(s-c)(s-a)\leq\dfrac{ab+bc+ca}{4}$.

Exercise 2.29. If a, b, c are the lengths of the sides of an acute triangle, prove that

$$\sum_{\text{cyclic}}\sqrt{a^2+b^2-c^2}\sqrt{a^2-b^2+c^2}\leq a^2+b^2+c^2,$$

where $\sum_{\text{cyclic}}$ stands for the sum over all cyclic permutations of (a,b,c).

Exercise 2.30. If a, b, c are the lengths of the sides of an acute triangle, prove that

$$\sum_{\text{cyclic}}\sqrt{a^2+b^2-c^2}\sqrt{a^2-b^2+c^2}\leq ab+bc+ca,$$

where $\sum_{\text{cyclic}}$ represents the sum over all cyclic permutations of (a,b,c).

2.3 The use of inequalities in the geometry of the triangle

A problem which shows the use of inequalities in the geometry of the triangle was introduced in the International Mathematical Olympiad in 1961; for this problem there are several proofs and its applications are very broad, as will be seen later on. Meanwhile, we present it here as an example.

Example 2.3.1. *If a, b and c are the lengths of the sides of a triangle with area (ABC), then $4\sqrt{3}(ABC) \leq a^2 + b^2 + c^2$.*

Since an equilateral triangle of side-length a has area equal to $\frac{\sqrt{3}}{4}a^2$, the equality in the example holds for this case; hence we will try to compare what happens in any triangle with what happens in an equilateral triangle of side length a.

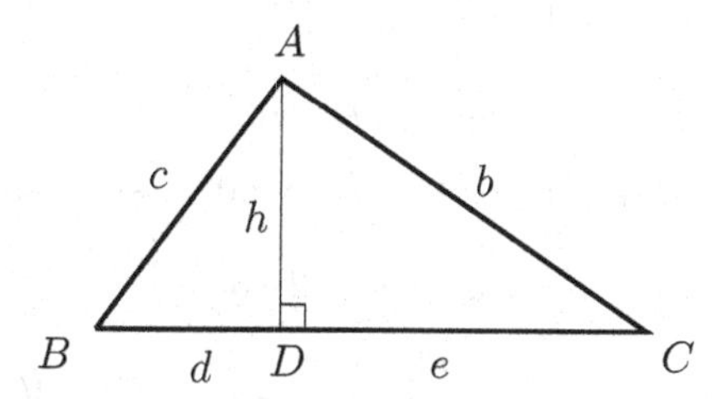

Let $BC = a$. If AD is the altitude of the triangle at A, its length h can be expressed as $h = \frac{\sqrt{3}}{2}a + y$, where y measures its difference in comparison with the length of the altitude of the equilateral triangle. We also set $d = \frac{a}{2} - x$ and $e = \frac{a}{2} + x$, where x can be interpreted as the difference that the projection of A on BC has with respect to the projection of A on BC in an equilateral triangle, which in this case is the midpoint of BC. We obtain

$$\begin{aligned}
a^2 + b^2 + c^2 - 4\sqrt{3}(ABC) &= a^2 + h^2 + \left(\frac{a}{2} + x\right)^2 + h^2 + \left(\frac{a}{2} - x\right)^2 - 4\sqrt{3}\frac{ah}{2} \\
&= \frac{3}{2}a^2 + 2h^2 + 2x^2 - 2\sqrt{3}\,a\left(\frac{\sqrt{3}}{2}a + y\right) \\
&= \frac{3}{2}a^2 + 2\left(\frac{\sqrt{3}}{2}a + y\right)^2 + 2x^2 - 3a^2 - 2\sqrt{3}ay \\
&= \frac{3}{2}a^2 + \frac{3}{2}a^2 + 2\sqrt{3}ay + 2y^2 + 2x^2 - 3a^2 - 2\sqrt{3}ay \\
&= 2(x^2 + y^2) \geq 0.
\end{aligned}$$

Moreover, the equality holds if and only if $x = y = 0$, that is, when the triangle is equilateral.

Let us give another proof for the previous example. Let ABC be a triangle, with side-lengths $a \geq b \geq c$, and let A' be a point such that $A'BC$ is an equilateral triangle with side-length a. If we take $d = AA'$, then d measures, in a manner, how far is ABC from being an equilateral triangle.

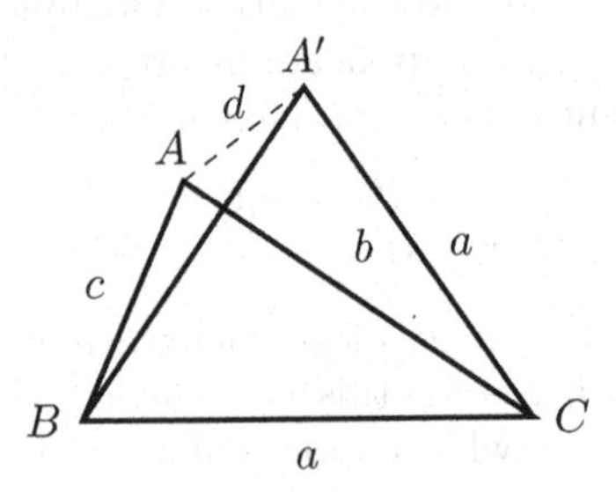

Using the cosine law we can deduce that

$$\begin{aligned} d^2 &= a^2 + c^2 - 2ac\cos(B - 60^\circ) \\ &= a^2 + c^2 - 2ac(\cos B \cos 60^\circ + \sin B \sin 60^\circ) \\ &= a^2 + c^2 - ac\cos B - 2\sqrt{3}\frac{ac\sin B}{2} \\ &= a^2 + c^2 - ac\left(\frac{a^2 + c^2 - b^2}{2ac}\right) - 2\sqrt{3}(ABC) \\ &= \frac{a^2 + b^2 + c^2}{2} - 2\sqrt{3}(ABC). \end{aligned}$$

But $d^2 \geq 0$, hence we can deduce that $4\sqrt{3}(ABC) \leq a^2 + b^2 + c^2$, which is what we wanted to prove. Moreover, the equality holds if $d = 0$, that is, if $A' = A$ or, equivalently, if ABC is equilateral.

It is quite common to find inequalities that involve elements of the triangle among mathematical olympiad problems. Some of them are based on the following inequality, which is valid for positive numbers a, b, c (see Exercise 1.36 of Section 1.3):

$$(a + b + c)\left(\frac{1}{a} + \frac{1}{b} + \frac{1}{c}\right) \geq 9. \tag{2.3}$$

Moreover, we recall that the equality holds if and only if $a = b = c$.

Another inequality, which has been very helpful to solve geometric-related problems, is Nesbitt's inequality (see Example 1.4.8 of Section 1.4). It states that for a, b, c positive numbers, we always have

$$\frac{a}{b+c} + \frac{b}{c+a} + \frac{c}{a+b} \geq \frac{3}{2}. \tag{2.4}$$

The previous inequality can be proved using inequality (2.3) as follows:

$$\begin{aligned}
\frac{a}{b+c}+\frac{b}{c+a}+\frac{c}{a+b} &= \frac{a+b+c}{b+c}+\frac{a+b+c}{c+a}+\frac{a+b+c}{a+b}-3\\
&= (a+b+c)\left(\frac{1}{b+c}+\frac{1}{c+a}+\frac{1}{a+b}\right)-3\\
&= \frac{1}{2}\left[(a+b)+(b+c)+(c+a)\right]\cdot\left(\frac{1}{b+c}+\frac{1}{c+a}+\frac{1}{a+b}\right)-3\\
&\geq \frac{9}{2}-3=\frac{3}{2}.
\end{aligned}$$

The equality holds if and only if $a+b=b+c=c+a$, or equivalently, if $a=b=c$.

Let us now observe some examples of geometric inequalities where such relationships are employed.

Example 2.3.2. *Let ABC be an equilateral triangle of side length a, let M be a point inside ABC and let D, E, F be the projections of M on the sides BC, CA and AB, respectively. Prove that*

(i) $\dfrac{1}{MD}+\dfrac{1}{ME}+\dfrac{1}{MF}\geq\dfrac{6\sqrt{3}}{a}$,

(ii) $\dfrac{1}{MD+ME}+\dfrac{1}{ME+MF}+\dfrac{1}{MF+MD}\geq\dfrac{3\sqrt{3}}{a}$.

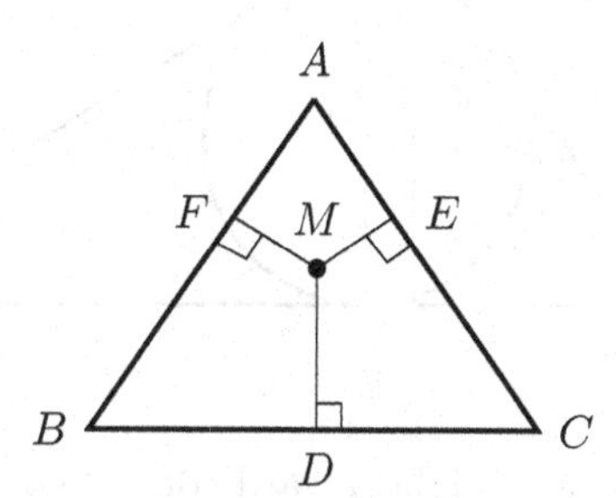

Let $x=MD$, $y=ME$ and $z=MF$. Remember that we denote the area of the triangle ABC as (ABC), then $(ABC)=(BCM)+(CAM)+(ABM)$, hence $ah=ax+ay+az$, where $h=\frac{\sqrt{3}}{2}a$ represents the length of the altitude of ABC. Therefore, $h=x+y+z$. (This result is known as Viviani's lemma; see Section 2.8). Using inequality (2.3) we can deduce that

$$h\left(\frac{1}{x}+\frac{1}{y}+\frac{1}{z}\right)\geq 9 \quad\text{and, after solving, that}\quad \frac{1}{x}+\frac{1}{y}+\frac{1}{z}\geq\frac{9}{h}=\frac{6\sqrt{3}}{a}.$$

To prove the second part, using inequality (2.3), we can establish that

$$(x+y+y+z+z+x)\left(\frac{1}{x+y}+\frac{1}{y+z}+\frac{1}{z+x}\right)\geq 9.$$

Therefore, $\frac{1}{x+y}+\frac{1}{y+z}+\frac{1}{z+x}\geq\frac{9}{2h}=\frac{3\sqrt{3}}{a}$.

Example 2.3.3. *If h_a, h_b and h_c are the lengths of the altitudes of the triangle ABC, whose incircle has center I and radius r, we have*

(i) $\dfrac{r}{h_a}+\dfrac{r}{h_b}+\dfrac{r}{h_c}=1,$

(ii) $h_a+h_b+h_c\geq 9r.$

In order to prove the first equation, observe that $\frac{r}{h_a}=\frac{r\cdot a}{h_a\cdot a}=\frac{(IBC)}{(ABC)}$. Similarly, $\frac{r}{h_b}=\frac{(ICA)}{(ABC)}$, $\frac{r}{h_c}=\frac{(IAB)}{(ABC)}$. Adding the three equations, we have that

$$\begin{aligned}\frac{r}{h_a}+\frac{r}{h_b}+\frac{r}{h_c}&=\frac{(IBC)}{(ABC)}+\frac{(ICA)}{(ABC)}+\frac{(IAB)}{(ABC)}\\&=\frac{(IBC)+(ICA)+(IAB)}{(ABC)}=1.\end{aligned}$$

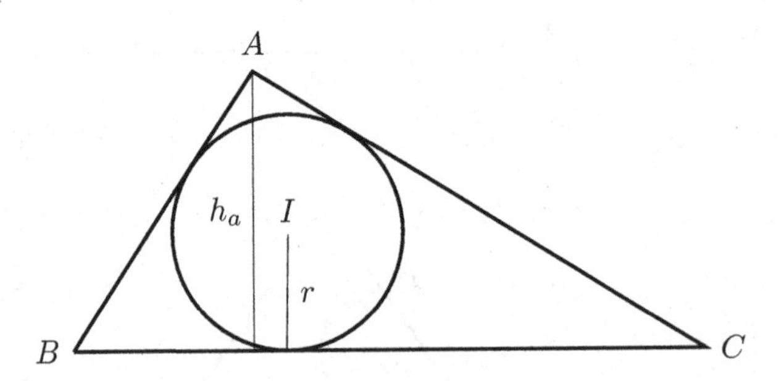

The desired inequality is a straightforward consequence of inequality (2.3), since $(h_a+h_b+h_c)\left(\frac{1}{h_a}+\frac{1}{h_b}+\frac{1}{h_c}\right)\cdot r\geq 9r$.

Example 2.3.4. *Let ABC be a triangle with altitudes AD, BE, CF and let H be its orthocenter. Prove that*

(i) $\dfrac{AD}{HD}+\dfrac{BE}{HE}+\dfrac{CF}{HF}\geq 9,$

(ii) $\dfrac{HD}{HA}+\dfrac{HE}{HB}+\dfrac{HF}{HC}\geq\dfrac{3}{2}.$

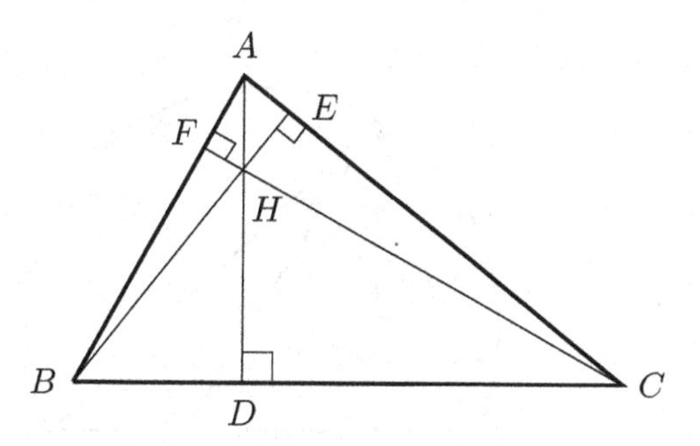

To prove part (i), consider $S = (ABC)$, $S_1 = (HBC)$, $S_2 = (HCA)$, $S_3 = (HAB)$. Since triangles ABC and HBC share the same base, their area ratio is equal to their altitude ratio, that is, $\frac{S_1}{S} = \frac{HD}{AD}$. Similarly, $\frac{S_2}{S} = \frac{HE}{BE}$ and $\frac{S_3}{S} = \frac{HF}{CF}$. Then, $\frac{HD}{AD} + \frac{HE}{BE} + \frac{HF}{CF} = 1$.

Using inequality (2.3) we can state that

$$\left(\frac{AD}{HD} + \frac{BE}{HE} + \frac{CF}{HF}\right)\left(\frac{HD}{AD} + \frac{HE}{BE} + \frac{HF}{CF}\right) \geq 9.$$

If we substitute the equality previously calculated, we get (i).

Moreover, the equality holds if and only if $\frac{HD}{AD} = \frac{HE}{BE} = \frac{HF}{CF} = \frac{1}{3}$, that is, if $S_1 = S_2 = S_3 = \frac{1}{3}S$. To prove the second part observe that $\frac{HD}{HA} = \frac{HD}{AD-HD} = \frac{S_1}{S-S_1} = \frac{S_1}{S_2+S_3}$, and similarly, $\frac{HE}{HB} = \frac{S_2}{S_3+S_1}$ and $\frac{HF}{HC} = \frac{S_3}{S_1+S_2}$, then using Nesbitt's inequality leads to $\frac{HD}{HA} + \frac{HE}{HB} + \frac{HF}{HC} \geq \frac{3}{2}$.

Example 2.3.5. (Korea, 1995) *Let ABC be a triangle and let L, M, N be points on BC, CA and AB, respectively. Let P, Q and R be the intersection points of the lines AL, BM and CN with the circumcircle of ABC, respectively. Prove that*

$$\frac{AL}{LP} + \frac{BM}{MQ} + \frac{CN}{NR} \geq 9.$$

Let A' be the midpoint of BC, let P' be the midpoint of the arc BC, let D and D' be the projections of A and P on BC, respectively.

It is clear that $\frac{AL}{LP} = \frac{AD}{PD'} \geq \frac{AD}{P'A'}$. Thus, the minimum value of $\frac{AL}{LP} + \frac{BM}{MQ} + \frac{CN}{NR}$ is attained when P, Q and R are the midpoints of the arcs BC, CA and AB. This happens when AL, BM and CN are the internal angle bisectors of the triangle ABC. Hence, without loss of generality, we will assume that AL, BM and CN are the internal angle bisectors of ABC. Since AL is an internal angle bisector, we have[6]

$$BL = \frac{ca}{b+c}, \quad LC = \frac{ba}{b+c} \quad \text{and} \quad AL^2 = bc\left(1 - \left(\frac{a}{b+c}\right)^2\right).$$

[6]See [6, pages 74 and 105] or [9, pages 10,11].

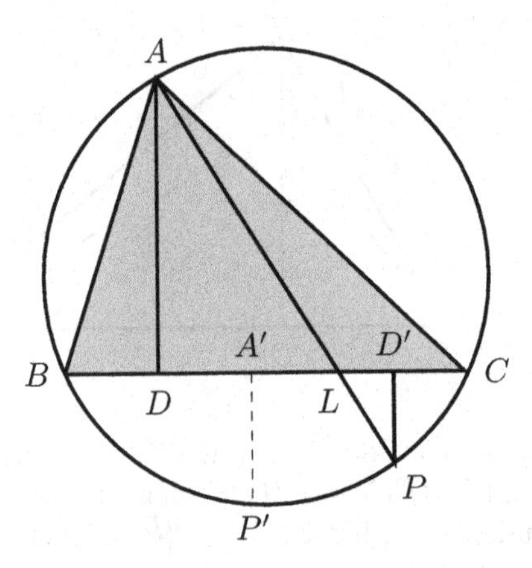

Moreover,

$$\frac{AL}{LP} = \frac{AL^2}{AL \cdot LP} = \frac{AL^2}{BL \cdot LC} = \frac{(bc)\left(1 - \left(\frac{a}{b+c}\right)^2\right)}{\frac{a^2bc}{(b+c)^2}} = \frac{(b+c)^2 - a^2}{a^2}.$$

Similarly, for the internal angle bisectors BM and CN, we have

$$\frac{BM}{MQ} = \frac{(c+a)^2 - b^2}{b^2} \quad \text{and} \quad \frac{CN}{NR} = \frac{(a+b)^2 - c^2}{c^2}.$$

Therefore,

$$\begin{aligned}\frac{AL}{LP} + \frac{BM}{MQ} + \frac{CN}{NR} &= \left(\frac{b+c}{a}\right)^2 + \left(\frac{c+a}{b}\right)^2 + \left(\frac{a+b}{c}\right)^2 - 3 \\ &\geq \frac{1}{3}\left(\frac{b+c}{a} + \frac{c+a}{b} + \frac{a+b}{c}\right)^2 - 3 \\ &\geq \frac{1}{3}(6)^2 - 3 = 9.\end{aligned}$$

The first inequality follows from the convexity of the function $f(x) = x^2$ and the second inequality from relations in the form $\frac{a}{b} + \frac{b}{a} \geq 2$. Observe that equality holds if and only if $a = b = c$.

Another way to finish the problem is the following:

$$\begin{aligned}&\left(\frac{b+c}{a}\right)^2 + \left(\frac{c+a}{b}\right)^2 + \left(\frac{a+b}{c}\right)^2 - 3 \\ &= \left(\frac{a^2}{b^2} + \frac{b^2}{a^2}\right) + \left(\frac{b^2}{c^2} + \frac{c^2}{b^2}\right) + \left(\frac{c^2}{a^2} + \frac{a^2}{c^2}\right) + 2\left(\frac{ab}{c^2} + \frac{bc}{a^2} + \frac{ca}{b^2}\right) - 3 \\ &\geq 2 \cdot 3 + 2 \cdot 3 - 3 = 9.\end{aligned}$$

Here we made use of the fact that $\frac{a^2}{b^2} + \frac{b^2}{a^2} \geq 2$ and that $\frac{ab}{c^2} + \frac{bc}{a^2} + \frac{ca}{b^2} \geq 3\sqrt[3]{\frac{(ab)(bc)(ca)}{a^2b^2c^2}} = 3$.

Example 2.3.6. (Shortlist IMO, 1997) *The lengths of the sides of the hexagon $ABCDEF$ satisfy $AB = BC$, $CD = DE$ and $EF = FA$. Prove that*

$$\frac{BC}{BE} + \frac{DE}{DA} + \frac{FA}{FC} \geq \frac{3}{2}.$$

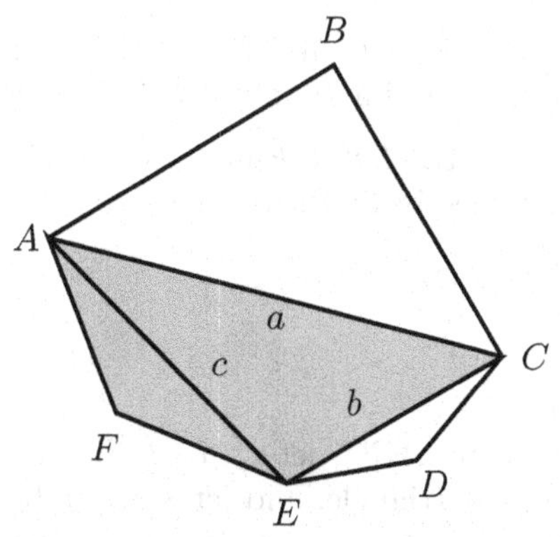

Set $a = AC$, $b = CE$ and $c = EA$. Ptolemy's inequality (see Exercise 2.11), applied to the quadrilateral $ACEF$, guarantees that $AE \cdot FC \leq FA \cdot CE + AC \cdot EF$. Since $EF = FA$, we have that $c \cdot FC \leq FA \cdot b + FA \cdot a$. Therefore,

$$\frac{FA}{FC} \geq \frac{c}{a+b}.$$

Similarly, we can deduce the inequalities

$$\frac{BC}{BE} \geq \frac{a}{b+c} \quad \text{and} \quad \frac{DE}{DA} \geq \frac{b}{c+a}.$$

Hence, $\frac{BC}{BE} + \frac{DE}{DA} + \frac{FA}{FC} \geq \frac{a}{b+c} + \frac{b}{c+a} + \frac{c}{a+b} \geq \frac{3}{2}$; the last inequality is Nesbitt's inequality.

Exercise 2.31. Let a, b, c be the lengths of the sides of a triangle, prove that:

(i) $\displaystyle \frac{a}{b+c-a} + \frac{b}{c+a-b} + \frac{c}{a+b-c} \geq 3,$

(ii) $\displaystyle \frac{b+c-a}{a} + \frac{c+a-b}{b} + \frac{a+b-c}{c} \geq 3.$

Exercise 2.32. Let AD, BE, CF be the altitudes of the triangle ABC and let PQ, PR, PS be the distances from a point P to the sides BC, CA, AB, respectively. Prove that

$$\frac{AD}{PQ} + \frac{BE}{PR} + \frac{CF}{PS} \geq 9.$$

Exercise 2.33. Through a point O inside a triangle of area S three lines are drawn in such a way that every side of the triangle intersects two of them. These lines divide the triangle into three triangles with common vertex O and areas S_1, S_2 and S_3, and three quadrilaterals. Prove that

(i) $\dfrac{1}{S_1}+\dfrac{1}{S_2}+\dfrac{1}{S_3}\geq\dfrac{9}{S}$,

(ii) $\dfrac{1}{S_1}+\dfrac{1}{S_2}+\dfrac{1}{S_3}\geq\dfrac{18}{S}$.

Exercise 2.34. The cevians AL, BM and CN of the triangle ABC concur in P. Prove that $\frac{AP}{PL}+\frac{BP}{PM}+\frac{CP}{PN}=6$ if and only if P is the centroid of the triangle.

Exercise 2.35. The altitudes AD, BE, CF intersect the circumcircle of the triangle ABC in D', E' and F', respectively. Prove that

(i) $\dfrac{AD}{DD'}+\dfrac{BE}{EE'}+\dfrac{CF}{FF'}\geq 9$,

(ii) $\dfrac{AD}{AD'}+\dfrac{BE}{BE'}+\dfrac{CE}{CF'}\geq\dfrac{9}{4}$.

Exercise 2.36. In the triangle ABC, let l_a, l_b, l_c be the lengths of the internal bisectors of the angles of the triangle, and let s and r be the semiperimeter and the inradius of ABC. Prove that

(i) $l_a l_b l_c\leq rs^2$,

(ii) $l_a l_b+l_b l_c+l_c l_a\leq s^2$,

(iii) $l_a^2+l_b^2+l_c^2\leq s^2$.

Exercise 2.37. Let ABC be a triangle and let M, N, P be arbitrary points on the line segments BC, CA, AB, respectively. Denote the lengths of the sides of the triangle by a, b, c and the circumradius by R. Prove that

$$\frac{bc}{AM}+\frac{ca}{BN}+\frac{ab}{CP}\leq 6R.$$

Exercise 2.38. Let ABC be a triangle with side-lengths a, b, c. Let m_a, m_b and m_c be the lengths of the medians from A, B and C, respectively. Prove that

$$\max\{a\,m_a, b\,m_b, c\,m_c\}\leq sR,$$

where R is the radius of the circumcircle and s is the semiperimeter.

2.4 Euler's inequality and some applications

Theorem 2.4.1 (Euler's theorem). *Given the triangle ABC, where O is the circumcenter, I the incenter, R the circumradius and r the inradius, then*

$$OI^2=R^2-2Rr.$$

Proof. Let us give a proof[7] that depends only on Pythagoras theorem and the fact that the circumcircle of the triangle BCI has center D, the midpoint of the arc[8] BC. For the proof we will use directed segments.

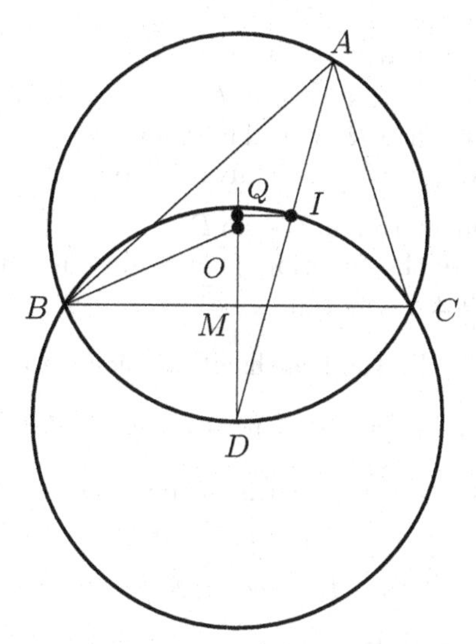

Let M be the midpoint of BC and let Q be the orthogonal projection of I on the radius OD. Then

$$\begin{aligned} OB^2 - OI^2 &= OB^2 - DB^2 + DI^2 - OI^2 \\ &= OM^2 - MD^2 + DQ^2 - QO^2 \\ &= (MO + DM)(MO - DM) + (DQ + QO)(DQ - QO) \\ &= DO(MO + MD + DQ + OQ) \\ &= R(2MQ) = 2Rr. \end{aligned}$$

Therefore $OI^2 = R^2 - 2Rr$. □

As a consequence of the last theorem we obtain the following inequality.

Theorem 2.4.2 (Euler's inequality). *$R \geq 2r$. Moreover, $R = 2r$ if and only if the triangle is equilateral.*[9]

---

[7]Another proof can be found in [6, page 122] or [9, page 29].

[8]The proof can be found in [6, observation 3.2.7, page 123] or [1, page 76].

[9]There are direct proofs for the inequality (that is, without having to use Euler's formula). One of them is the following: the nine-point circle of a triangle is the circumcircle of the medial triangle $A'B'C'$. Because this triangle is similar to ABC with ratio 2:1, we can deduce that the radius of the nine-point circle is $\frac{R}{2}$. Clearly, a circle that intersects the three sides of a triangle must have a greater radius than the radius of the incircle, therefore $\frac{R}{2} \geq r$.

Theorem 2.4.3. *In a triangle ABC, with circumradius R, inradius r and semiperimeter s, it happens that*

$$r \leq \frac{s}{3\sqrt{3}} \leq \frac{R}{2}.$$

Proof. We will use the fact that[10] $(ABC) = \frac{abc}{4R} = sr$. Using the *AM-GM* inequality, we can deduce that $2s = a+b+c \geq 3\sqrt[3]{abc} = 3\sqrt[3]{4Rrs}$. Thus, $8s^3 \geq 27(4Rrs) \geq 27(8r^2s)$, since $R \geq 2r$. Therefore, $s \geq 3\sqrt{3}r$.

The second inequality, $\frac{s}{3\sqrt{3}} \leq \frac{R}{2}$, is equivalent to $a+b+c \leq 3\sqrt{3}R$. But using the sine law, this is equivalent to $\sin A + \sin B + \sin C \leq \frac{3\sqrt{3}}{2}$. Observe that the last inequality holds because the function $f(x) = \sin x$ is concave on $[0, \pi]$, thus $\frac{\sin A + \sin B + \sin C}{3} \leq \sin\left(\frac{A+B+C}{3}\right) = \sin 60° = \frac{\sqrt{3}}{2}$. □

Exercise 2.39. Let a, b and c be the lengths of the sides of a triangle, prove that

$$(a+b-c)(b+c-a)(c+a-b) \leq abc.$$

Exercise 2.40. Let a, b and c be the lengths of the sides of a triangle, prove that

$$\frac{1}{ab} + \frac{1}{bc} + \frac{1}{ca} \geq \frac{1}{R^2},$$

where R denotes the circumradius.

Exercise 2.41. Let A, B and C be the measurements of the angles in each of the vertices of the triangle ABC, prove that

$$\frac{1}{\sin A \sin B} + \frac{1}{\sin B \sin C} + \frac{1}{\sin C \sin A} \geq 4.$$

Exercise 2.42. Let A, B and C be the measurements of the angles in each of the vertices of the triangle ABC, prove that

$$\left(\sin\frac{A}{2}\right)\left(\sin\frac{B}{2}\right)\left(\sin\frac{C}{2}\right) \leq \frac{1}{8}.$$

Exercise 2.43. Let ABC be a triangle. Call A, B and C the angles in the vertices A, B and C, respectively. Let a, b and c be the lengths of the sides of the triangle and let R be the radius of the circumcircle. Prove that

$$\left(\frac{2A}{\pi}\right)^{\frac{1}{a}}\left(\frac{2B}{\pi}\right)^{\frac{1}{b}}\left(\frac{2C}{\pi}\right)^{\frac{1}{c}} \leq \left(\frac{2}{3}\right)^{\frac{\sqrt{3}}{R}}.$$

Theorem 2.4.4 (Leibniz's theorem). *In a triangle ABC with sides of length a, b and c, and with circumcenter O, centroid G and circumradius R, the following holds:*

$$OG^2 = R^2 - \frac{1}{9}\left(a^2+b^2+c^2\right).$$

[10] See [6, page 97] or [9, page 13].

Proof. Let us use Stewart's theorem which states[11] that if L is a point on the side BC of a triangle ABC and if $AL = l$, $BL = m$, $LC = n$, then $a\left(l^2 + mn\right) = b^2m + c^2n$.

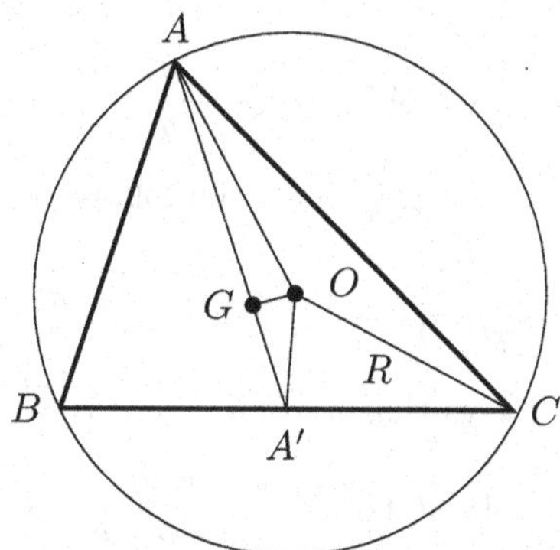

Applying Stewart's theorem to the triangle OAA' to find the length of OG, where A' is the midpoint of BC, we get

$$AA'\left(OG^2 + AG \cdot GA'\right) = A'O^2 \cdot AG + AO^2 \cdot GA'.$$

Since

$$AO = R, \quad AG = \frac{2}{3}AA' \quad \text{and} \quad GA' = \frac{1}{3}AA',$$

substituting we get

$$OG^2 + \frac{2}{9}(A'A)^2 = A'O^2 \cdot \frac{2}{3} + R^2 \cdot \frac{1}{3}.$$

On the other hand[12], since $(A'A)^2 = \frac{2(b^2+c^2)-a^2}{4}$ and $A'O^2 = R^2 - \frac{a^2}{4}$, we can deduce that

$$\begin{aligned} OG^2 &= \left(R^2 - \frac{a^2}{4}\right)\frac{2}{3} + \frac{1}{3}R^2 - \frac{2}{9}\left(\frac{2\left(b^2+c^2\right) - a^2}{4}\right) \\ &= R^2 - \frac{a^2}{6} - \frac{2\left(b^2+c^2\right) - a^2}{18} \\ &= R^2 - \frac{a^2+b^2+c^2}{9}. \end{aligned}$$

□

One consequence of the last theorem is the following inequality.

Theorem 2.4.5 (Leibniz's nequality). *In a triangle ABC with side-lengths a, b and c, with circumradius R, the following holds:*

$$9R^2 \geq a^2 + b^2 + c^2.$$

[11] For a proof see [6, page 96] or [9, page 6].

[12] See [6, page 83] or [9, page 10].

Moreover, equality holds if and only if $O = G$, that is, when the triangle is equilateral.

Example 2.4.6. *In a triangle ABC with sides of length a, b and c, it follows that*

$$4\sqrt{3}(ABC) \leq \frac{9abc}{a+b+c}.$$

Using that $4R\,(ABC) = abc$, we have the following equivalences:

$$9R^2 \geq a^2+b^2+c^2 \Leftrightarrow \frac{a^2b^2c^2}{16(ABC)^2} \geq \frac{a^2+b^2+c^2}{9} \Leftrightarrow 4(ABC) \leq \frac{3abc}{\sqrt{a^2+b^2+c^2}}.$$

Cauchy-Schwarz inequality says that $a+b+c \leq \sqrt{3}\sqrt{a^2+b^2+c^2}$, hence

$$4\sqrt{3}(ABC) \leq \frac{9abc}{a+b+c}.$$

Exercise 2.44. Let A, B and C be the measurements of the angles in each of the vertices of the triangle ABC, prove that

$$\sin^2 A + \sin^2 B + \sin^2 C \leq \frac{9}{4}.$$

Exercise 2.45. Let a, b and c be the lengths of the sides of a triangle, prove that

$$4\sqrt{3}(ABC) \leq 3\sqrt[3]{a^2b^2c^2}.$$

Exercise 2.46. Suppose that the incircle of ABC is tangent to the sides BC, CA, AB, at D, E, F, respectively. Prove that

$$EF^2 + FD^2 + DE^2 \leq \frac{s^2}{3},$$

where s is the semiperimeter of ABC.

Exercise 2.47. Let a, b, c be the lenghts of the sides of a triangle ABC and let h_a, h_b, h_c be the lenghts of the altitudes over BC, CA, AB, respectively. Prove that

$$\frac{a^2}{h_bh_c} + \frac{b^2}{h_ch_a} + \frac{c^2}{h_ah_b} \geq 4.$$

2.5 Symmetric functions of a, b and c

The lengths of the sides a, b and c of a triangle have a very close relationship with s, r and R, the semiperimeter, the inradius and the circumradius of the triangle, respectively. The relationships that are most commonly used are

$$a+b+c = 2s, \tag{2.5}$$

$$ab+bc+ca = s^2+r^2+4rR, \tag{2.6}$$

$$abc = 4Rrs. \tag{2.7}$$

The first is the definition of s and the third follows from the fact that the area of the triangle is $\frac{abc}{4R} = rs$. Using Heron's formula for the area of a triangle, we have the relationship $s(s-a)(s-b)(s-c) = r^2s^2$, hence

$$s^3 - (a+b+c)s^2 + (ab+bc+ca)s - abc = r^2 s.$$

If we substitute (2.5) and (2.7) in this equality, after simplifying we get that

$$ab + bc + ca = s^2 + r^2 + 4Rr.$$

Now, since any symmetric polynomial in a, b and c can be expressed as a polynomial in terms of $(a+b+c)$, $(ab+bc+ca)$ and (abc), it can also be expressed as a polynomial in s, r and R. For instance,

$$\begin{aligned}
a^2+b^2+c^2 &= (a+b+c)^2 - 2(ab+bc+ca) = 2\left(s^2 - r^2 - 4Rr\right), \\
a^3+b^3+c^3 &= (a+b+c)^3 - 3(a+b+c)(ab+bc+ca) + 3abc \\
&= 2\left(s^3 - 3r^2 s - 6Rrs\right).
\end{aligned}$$

These transformations help to solve different problems, as will be seen later on.

Lemma 2.5.1. *If A, B and C are the measurements of the angles within each of the vertices of the triangle ABC, we have that* $\cos A + \cos B + \cos C = \frac{r}{R} + 1$.

Proof.

$$\begin{aligned}
\cos A + \cos B + \cos C &= \frac{b^2+c^2-a^2}{2bc} + \frac{c^2+a^2-b^2}{2ca} + \frac{a^2+b^2-c^2}{2ab} \\
&= \frac{a\left(b^2+c^2\right) + b\left(c^2+a^2\right) + c\left(a^2+b^2\right) - \left(a^3+b^3+c^3\right)}{2abc} \\
&= \frac{(a+b+c)\left(a^2+b^2+c^2\right) - 2(a^3+b^3+c^3)}{2abc} \\
&= \frac{4s\left(s^2-r^2-4Rr\right) - 4\left(s^3-3r^2s-6Rrs\right)}{8Rrs} \\
&= \frac{\left(s^2-r^2-4Rr\right) - \left(s^2-3r^2-6Rr\right)}{2Rr} \\
&= \frac{2r^2+2Rr}{2Rr} = \frac{r}{R} + 1.
\end{aligned}$$

□

Example 2.5.2. *If A, B and C are the measurements of the angles in each of the vertices of the triangle ABC, we have that* $\cos A + \cos B + \cos C \leq \frac{3}{2}$.

Lemma 2.5.1 guarantees that $\cos A + \cos B + \cos C = \frac{r}{R} + 1$, and using Euler's inequality, $R \geq 2r$, we get the result.

We can give another direct proof. Observe that,

$$a(b^2+c^2-a^2)+b(c^2+a^2-b^2)+c(a^2+b^2-c^2) = (b+c-a)(c+a-b)(a+b-c)+2abc.$$

Then,

$$\begin{aligned}\cos A+\cos B+\cos C &= \frac{b^2+c^2-a^2}{2bc}+\frac{c^2+a^2-b^2}{2ca}+\frac{a^2+b^2-c^2}{2ab}\\ &= \frac{(b+c-a)(c+a-b)(a+b-c)}{2abc}+1,\end{aligned}$$

and since $(b+c-a)(c+a-b)(a+b-c) \leq abc$, we have the result.

Example 2.5.3. (IMO, 1991) *Let ABC be a triangle, let I be its incenter and let L, M, N be the intersections of the internal angle bisectors of A, B, C with BC, CA, AB, respectively. Prove that $\frac{1}{4} < \frac{AI}{AL}\frac{BI}{BM}\frac{CI}{CN} \leq \frac{8}{27}$.*

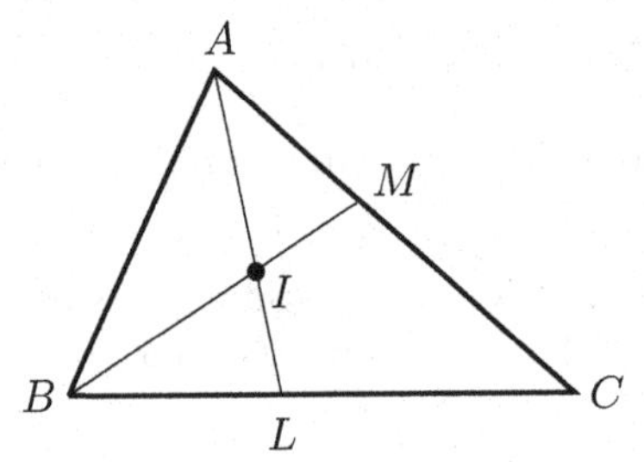

Using the angle bisector theorem $\frac{BL}{LC} = \frac{AB}{CA} = \frac{c}{b}$ and the fact that $BL + LC = a$, we can deduce that $BL = \frac{ac}{b+c}$ and $LC = \frac{ab}{b+c}$. Again, the angle bisector theorem applied to the internal angle bisector BI of the angle $\angle ABL$ gives us $\frac{IL}{AI} = \frac{BL}{AB} = \frac{ac}{(b+c)c} = \frac{a}{b+c}$. Hence,

$$\frac{AL}{AI} = \frac{AI+IL}{AI} = 1+\frac{IL}{AI} = 1+\frac{a}{b+c} = \frac{a+b+c}{b+c}.$$

Then, $\frac{AI}{AL} = \frac{b+c}{a+b+c}$.[13] Similarly, $\frac{BI}{BM} = \frac{c+a}{a+b+c}$ and $\frac{CI}{CN} = \frac{a+b}{a+b+c}$. Therefore, the inequality that we have to prove in terms of a, b and c is

$$\frac{1}{4} < \frac{(b+c)(c+a)(a+b)}{(a+b+c)^3} \leq \frac{8}{27}.$$

The *AM-GM* inequality guarantees that

$$(b+c)(c+a)(a+b) \leq \left(\frac{(b+c)+(c+a)+(a+b)}{3}\right)^3 = \frac{8}{27}(a+b+c)^3,$$

[13] Another way to prove the identity is as follows. Consider $\alpha = (ABI)$, $\beta = (BCI)$ and $\gamma = (CAI)$. It is clear that $\frac{AI}{AL} = \frac{\alpha+\gamma}{\alpha+\beta+\gamma} = \frac{r(c+b)}{r(a+c+b)} = \frac{c+b}{a+c+b}$.

hence the inequality on the right-hand side is now evident.

To prove the inequality on the left-hand side, first note that

$$\frac{(b+c)(c+a)(a+b)}{(a+b+c)^3} = \frac{(a+b+c)(ab+bc+ca)-abc}{(a+b+c)^3}.$$

Substitute above, using equations (2.5), (2.6) and (2.7), to get

$$\begin{aligned}\frac{(b+c)(c+a)(a+b)}{(a+b+c)^3} &= \frac{2s(s^2+r^2+4Rr)-4Rrs}{8s^3}\\ &= \frac{2s^3+2sr^2+4Rrs}{8s^3} = \frac{1}{4}+\frac{2r^2+4Rr}{8s^2} > \frac{1}{4}.\end{aligned}$$

We can also use the Ravi transformation $a = y+z$, $b = z+x$, $c = x+y$, to reach the final result in the following way:

$$\begin{aligned}\frac{(b+c)(c+a)(a+b)}{(a+b+c)^3} &= \frac{(x+y+z+x)(x+y+z+y)(x+y+z+z)}{8(x+y+z)^3}\\ &= \frac{1}{8}\left(1+\frac{x}{x+y+z}\right)\left(1+\frac{y}{x+y+z}\right)\left(1+\frac{z}{x+y+z}\right)\\ &= \frac{1}{8}\left(1+\frac{x+y+z}{x+y+z}+\frac{xy+yz+zx}{x+y+z}+\frac{xyz}{x+y+z}\right) > \frac{1}{4}.\end{aligned}$$

Exercise 2.48. Let A, B and C be the values of the angles in each one of the vertices of the triangle ABC, prove that

$$\sin^2\frac{A}{2}+\sin^2\frac{B}{2}+\sin^2\frac{C}{2} \geq \frac{3}{4}.$$

Exercise 2.49. Let a, b and c be the lengths of the sides of a triangle. Using the tools we have studied in this section, prove that

$$4\sqrt{3}(ABC) \leq \frac{9abc}{a+b+c}.$$

Exercise 2.50. Let a, b and c be the lengths of the sides of a triangle. Using the tools presented in this section, prove that

$$4\sqrt{3}(ABC) \leq 3\sqrt[3]{a^2b^2c^2}.$$

Exercise 2.51. (IMO, 1961) Let a, b and c be the lengths of the sides of a triangle, prove that

$$4\sqrt{3}(ABC) \leq a^2+b^2+c^2.$$

Exercise 2.52. Let a, b and c be the lengths of the sides of a triangle, prove that

$$4\sqrt{3}(ABC) \leq a^2+b^2+c^2-(a-b)^2-(b-c)^2-(c-a)^2.$$

Exercise 2.53. Let a, b and c be the lengths of the sides of a triangle, prove that

$$4\sqrt{3}(ABC) \leq ab + bc + ca.$$

Exercise 2.54. Let a, b and c be the lengths of the sides of a triangle, prove that

$$4\sqrt{3}(ABC) \leq \frac{3(a+b+c)abc}{ab+bc+ca}.$$

Exercise 2.55. Let a, b and c be the lengths of the sides of a triangle. If $a+b+c=1$, prove that

$$a^2 + b^2 + c^2 + 4abc < \frac{1}{2}.$$

Exercise 2.56. Let a, b and c be the lengths of the sides of a triangle, let R and r be the circumradius and the inradius, respectively, prove that

$$\frac{(b+c-a)(c+a-b)(a+b-c)}{abc} = \frac{2r}{R}.$$

Exercise 2.57. Let a, b and c be the lengths of the sides of a triangle and let R be the circumradius, prove that

$$3\sqrt{3}R \leq \frac{a^2}{b+c-a} + \frac{b^2}{c+a-b} + \frac{c^2}{a+b-c}.$$

Exercise 2.58. Let a, b and c be the lengths of the sides of a triangle. Set $x = \frac{b+c-a}{2}$, $y = \frac{c+a-b}{2}$ and $z = \frac{a+b-c}{2}$. If $\tau_1 = x + y + z$, $\tau_2 = xy + yz + zx$ and $\tau_3 = xyz$, verify the following relationships.

(1) $(a-b)^2 + (b-c)^2 + (c-a)^2 = (x-y)^2 + (y-z)^2 + (z-x)^2 = 2(\tau_1^2 - 3\tau_2)$.

(2) $a + b + c = 2\tau_1$.

(3) $a^2 + b^2 + c^2 = 2\tau_1^2 - 2\tau_2$.

(4) $ab + bc + ca = \tau_1^2 + \tau_2$.

(5) $abc = \tau_1\tau_2 - \tau_3$.

(6) $16(ABC)^2 = 2(a^2b^2 + b^2c^2 + c^2a^2) - (a^4 + b^4 + c^4) = 16r^2s^2 = 16\tau_1\tau_3$.

(7) $R = \dfrac{\tau_1\tau_2 - \tau_3}{4\sqrt{\tau_1\tau_3}}$.

(8) $r = \sqrt{\dfrac{\tau_3}{\tau_1}}$.

(9) $\tau_1 = s$, $\tau_2 = r(4R + r)$, $\tau_3 = r^2 s$.

2.6 Inequalities with areas and perimeters

We begin this section with the following example.

Example 2.6.1. (Austria–Poland, 1985) *If $ABCD$ is a convex quadrilateral of area 1, then*

$$AB + BC + CD + DA + AC + BD \geq 4 + \sqrt{8}.$$

Set $a = AB$, $b = BC$, $c = CD$, $d = DA$, $e = AC$ and $f = BD$. The area of the quadrilateral $ABCD$ is $(ABCD) = \frac{ef\sin\theta}{2}$, where θ is the angle between the diagonals, which makes it clear that $1 = \frac{ef\sin\theta}{2} \leq \frac{ef}{2}$.

Since $(ABC) = \frac{ab\sin B}{2} \leq \frac{ab}{2}$ and $(CDA) = \frac{cd\sin D}{2} \leq \frac{cd}{2}$, we can deduce that $1 = (ABCD) \leq \frac{ab+cd}{2}$. Similarly, $1 = (ABCD) \leq \frac{bc+da}{2}$. These two inequalities imply that $ab + bc + cd + da \geq 4$.

Finally, since $(e+f)^2 = 4ef + (e-f)^2 \geq 4ef \geq 8$ and $(a+b+c+d)^2 = 4(a+c)(b+d) + ((a+c)-(b+d))^2 \geq 4(a+c)(b+d) = 4(ab+bc+cd+da) \geq 16$, we can deduce that $a+b+c+d+e+f \geq 4+\sqrt{8}$.

Example 2.6.2. (Iberoamerican, 1992) *Using the triangle ABC, construct a hexagon H with vertices A_1, A_2, B_1, B_2, C_1, C_2 as shown in the figure. Show that the area of the hexagon H is at least thirteen times the area of the triangle ABC.*

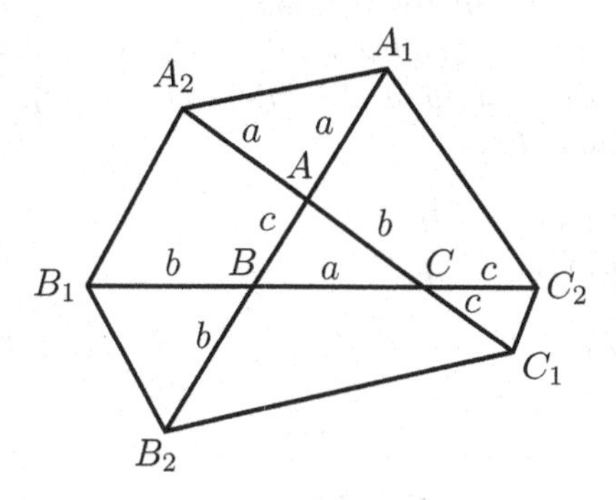

It is clear, using the area formula $(ABC) = \frac{ab\sin C}{2}$, that

$$\begin{aligned}(A_1A_2B_1B_2C_1C_2) =& (A_1BC_2) + (A_2CB_1) + (B_2AC_1) + (AA_1A_2) \\ &+ (BB_1B_2) + (CC_1C_2) - 2(ABC) \\ =& \frac{(c+a)^2\sin B}{2} + \frac{(a+b)^2\sin C}{2} + \frac{(b+c)^2\sin A}{2} \\ &+ \frac{a^2\sin A}{2} + \frac{b^2\sin B}{2} + \frac{c^2\sin C}{2} - 2(ABC)\end{aligned}$$

$$= \frac{(a^2+b^2+c^2)(\sin A + \sin B + \sin C)}{2} + ca \sin B$$

$$+ ab \sin C + bc \sin A - 2(ABC)$$

$$= \frac{(a^2+b^2+c^2)(\sin A + \sin B + \sin C)}{2} + 4(ABC).$$

Therefore, $(A_1A_2B_1B_2C_1C_2) \geq 13(ABC)$ if and only if

$$\frac{(a^2+b^2+c^2)(\sin A + \sin B + \sin C)}{2} \geq 9(ABC) = \frac{9abc}{4R}.$$

Using the sine law, $\frac{\sin A}{a} = \frac{1}{2R}$, we can prove that the inequality is true if and only if $\frac{(a^2+b^2+c^2)(a+b+c)}{4R} \geq \frac{9abc}{4R}$, that is,

$$(a^2+b^2+c^2)(a+b+c) \geq 9abc.$$

The last inequality can be deduced from the *AM-GM* inequality, from the rearrangement inequality or by using Tchebychev's inequality. Moreover, the equality holds only in the case $a = b = c$.

Example 2.6.3. (China, 1988 and 1993) *Consider two concentric circles of radii R and R_1 ($R_1 > R$) and a convex quadrilateral $ABCD$ inscribed in the small circle. The extensions of AB, BC, CD and DA intersect the large circle at C_1, D_1, A_1 and B_1, respectively. Show that*

(i) $\dfrac{\text{perimeter of } A_1B_1C_1D_1}{\text{perimeter of } ABCD} \geq \dfrac{R_1}{R}$;

(ii) $\dfrac{(A_1B_1C_1D_1)}{(ABCD)} \geq \left(\dfrac{R_1}{R}\right)^2$.

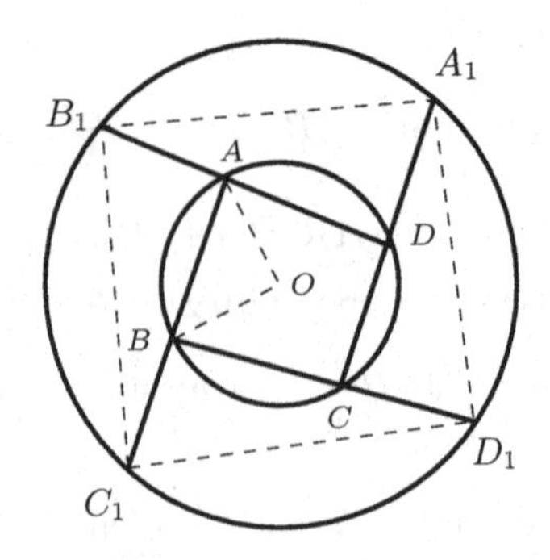

To prove (i), we use Ptolemy's inequality (see Exercise 2.11) applied to the quadrilaterals OAB_1C_1, OBC_1D_1, OCD_1A_1 and ODA_1B_1, which implies that

$$\begin{aligned} AC_1 \cdot R_1 &\leq B_1C_1 \cdot R + AB_1 \cdot R_1, \\ BD_1 \cdot R_1 &\leq C_1D_1 \cdot R + BC_1 \cdot R_1, \\ CA_1 \cdot R_1 &\leq D_1A_1 \cdot R + CD_1 \cdot R_1, \\ DB_1 \cdot R_1 &\leq A_1B_1 \cdot R + DA_1 \cdot R_1. \end{aligned} \tag{2.8}$$

Then, when we add these inequalities together and write AC_1, BD_1, CA_1 and DB_1, and express them as $AB + BC_1$, $BC + CD_1$, $CD + DA_1$ and $DA + AB_1$, respectively, we get

$$\begin{aligned} &R_1 \cdot \text{perimeter}\,(ABCD) + R_1(BC_1 + CD_1 + DA_1 + AB_1) \\ &\quad \leq R \cdot \text{perimeter}\,(A_1B_1C_1D_1) + R_1(AB_1 + BC_1 + CD_1 + DA_1). \end{aligned}$$

Therefore,

$$\frac{\text{perimeter}\,(A_1B_1C_1D_1)}{\text{perimeter}\,(ABCD)} \geq \frac{R_1}{R}.$$

To prove (ii), we use the fact that $(ABCD) = \frac{ad \sin A + bc \sin A}{2} = \frac{\sin A}{2}(ad+bc)$ and also that $(ABCD) = \frac{ab \sin B + cd \sin B}{2} = \frac{\sin B}{2}(ab+cd)$, where $A = \angle DAB$ and $B = \angle ABC$.

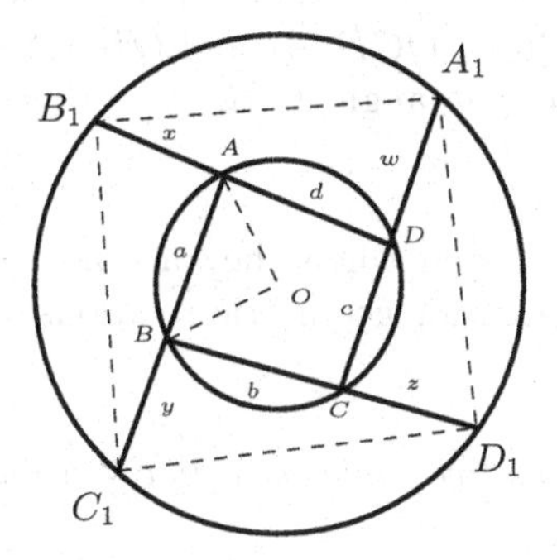

Since $(AB_1C_1) = \frac{x(a+y)\sin(180^\circ - A)}{2} = \frac{x(a+y)\sin A}{2}$, we can produce the identity $\frac{(AB_1C_1)}{(ABCD)} = \frac{x(a+y)}{ad+bc}$. Similarly, $\frac{(BC_1D_1)}{(ABCD)} = \frac{y(b+z)}{ab+cd}$, $\frac{(CD_1A_1)}{(ABCD)} = \frac{z(c+w)}{ad+bc}$, $\frac{(DA_1B_1)}{(ABCD)} = \frac{w(d+x)}{ab+cd}$. Then,

$$\frac{(A_1B_1C_1D_1)}{(ABCD)} = 1 + \frac{x(a+y) + z(w+c)}{ad+bc} + \frac{y(b+z) + w(d+x)}{ab+cd}.$$

The power of a point in the larger circle with respect to the small circle is equal to $R_1^2 - R^2$. In particular, the power of A_1, B_1, C_1 and D_1 is the same. On the other hand, we know that these powers are $w(w+c)$, $x(x+d)$, $y(y+a)$ and $z(z+b)$, respectively.

Substituting this in the previous equation implies that the area ratio is

$$\frac{(A_1B_1C_1D_1)}{(ABCD)} = 1+(R_1^2-R^2)\left[\frac{x}{y(ad+bc)}+\frac{z}{w(ad+bc)}+\frac{y}{z(ab+cd)}+\frac{w}{x(ab+cd)}\right].$$

Using the *AM-GM* inequality allows us to deduce that

$$\frac{(A_1B_1C_1D_1)}{(ABCD)} \geq 1 + \frac{4(R_1^2-R^2)}{\sqrt{(ad+bc)(ab+cd)}}.$$

Since $2\sqrt{(ad+bc)(ab+cd)} \leq ad+bc+ab+cd = (a+c)(b+d) \leq \frac{1}{4}(a+b+c+d)^2 \leq \frac{(4\sqrt{2}R)^2}{4} = 8R^2$, the first two inequalities follow from the *AM-GM* inequality, and the last one follows from the fact that, of all the quadrilaterals inscribed in a circle, the square has the largest perimeter. Thus

$$\frac{(A_1B_1C_1D_1)}{(ABCD)} \geq 1 + \frac{4(R_1^2-R^2)}{4R^2} = \left(\frac{R_1}{R}\right)^2.$$

Moreover, the equalities hold when $ABCD$ is a square and only in this case. Since in order to reduce inequalities (2.8) to identities, it must be the case that the four quadrilaterals OAB_1C_1, OBC_1D_1, OCD_1A_1 and ODA_1B_1 are cyclic. Thus, OA is an internal angle bisector of the angle BAD, and the same happens for OB, OC and OD.

There are problems that, even when they are not presented in a geometric form, they invite us to search for geometric relationships, as in the following example.

Example 2.6.4. *If a, b, c are positive numbers with $c < a$ and $c < b$, we can deduce that $\sqrt{c(a-c)} + \sqrt{c(b-c)} \leq \sqrt{ab}$.*

Consider the isosceles triangles ABC and ACD, both sharing the common side AC of length $2\sqrt{c}$; we take the first triangle as having equal sides $AB = BC$ of length $\sqrt{a}$ and the second one satisfying $CD = DA$ with length $\sqrt{b}$.

The area of the quadrilateral $ABCD$ is, on the one hand,

$$(ABCD) = (ABC) + (ACD) = \sqrt{c(a-c)} + \sqrt{b(b-c)};$$

and on the other hand, $(ABCD) = 2(ABD) = \frac{2\sqrt{ab}\sin\angle BAD}{2}$.

This last procedure for calculating the area clearly proves that $(ABCD) \leq \sqrt{ab}$, and thus the result is obtained.

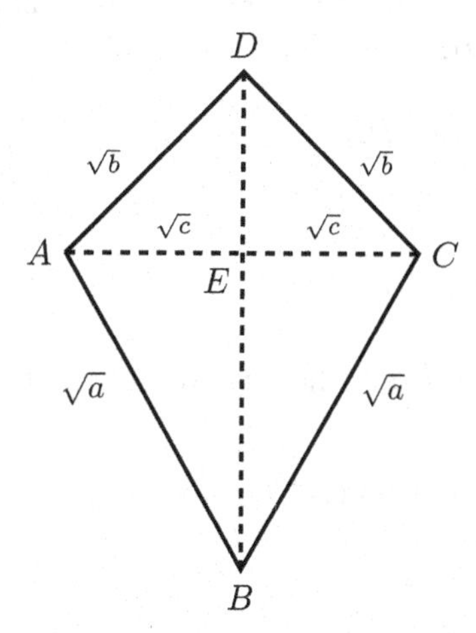

Another solution is as follows. Since AC and BD are perpendiculars, Pythagoras theorem implies that $DE = \sqrt{b-c}$ and $EB = \sqrt{a-c}$. By Ptolemy's inequality (see Exercise 2.11), $(\sqrt{b-c} + \sqrt{a-c}) \cdot (2\sqrt{c}) \leq \sqrt{a}\sqrt{b} + \sqrt{a}\sqrt{b}$ and then the result.

Exercise 2.59. On every side of a square with sides measuring 1, choose one point. The four points will form a quadrilateral of sides of length a, b, c and d, prove that

(i) $2 \leq a^2 + b^2 + c^2 + d^2 \leq 4$,

(ii) $2\sqrt{2} \leq a + b + c + d \leq 4$.

Exercise 2.60. On each side of a regular hexagon with sides measuring 1, we choose one point. The six points form a hexagon of perimeter h. Prove that $3\sqrt{3} \leq h \leq 6$.

Exercise 2.61. Consider the three lines tangent to the incircle of a triangle ABC which are parallel to the sides of the triangle; these, together with the sides of the triangle, form a hexagon T. Prove that

$$\text{the perimeter of } T \leq \frac{2}{3} \text{ the perimeter of } (ABC).$$

Exercise 2.62. Find the radius of the circle of maximum area that can be covered using three circles with radius 1.

Exercise 2.63. Find the radius of the circle of maximum area that can be covered using three circles with radii r_1, r_2 and r_3.

Exercise 2.64. Two disjoint squares are located inside a square of side 1. If the lengths of the sides of the two squares are a and b, prove that $a + b \leq 1$.

Exercise 2.65. A convex quadrilateral is inscribed in a circumference of radius 1, in such a way that one of its sides is a diameter and the other sides have lengths a, b and c. Prove that $abc \leq 1$.

Exercise 2.66. Let $ABCDE$ be a convex pentagon such that the areas of the triangles ABC, BCD, CDE, DEA and EAB are equal. Prove that

(i) $\dfrac{(ABCDE)}{4} < (ABC) < \dfrac{(ABCDE)}{3}$,

(ii) $(ABCDE) = \dfrac{5+\sqrt{5}}{2}(ABC)$.

Exercise 2.67. If AD, BE and CF are the altitudes of the triangle ABC, prove that

$$\text{perimeter}\,(DEF) \leq s,$$

where s is the semiperimeter.

Exercise 2.68. The lengths of the internal angle bisectors of a triangle are at most 1, show that the area of such a triangle is at most $\frac{\sqrt{3}}{3}$.

Exercise 2.69. If a, b, c, d are the lengths of the sides of a convex quadrilateral, show that

(i) $(ABCD) \leq \dfrac{ab+cd}{2}$,

(ii) $(ABCD) \leq \dfrac{ac+bd}{2}$ and

(iii) $(ABCD) \leq \left(\dfrac{a+c}{2}\right)\left(\dfrac{b+d}{2}\right)$.

2.7 Erdős-Mordell Theorem

Theorem 2.7.1 (Pappus's theorem). *Let ABC be a triangle, $AA'B'B$ and $CC'A''A$ two parallelograms constructed on AC and AB such that both either are inside or outside the triangle. Let P be the intersection of $B'A'$ with $C'A''$. Construct another parallelogram $BP'P''C$ on BC such that BP' is parallel to AP and of the same length. Thus, we will have the following relationships between the areas:*

$$(BP'P''C) = (AA'B'B) + (CC'A''A).$$

Proof. See the picture on the next page. □

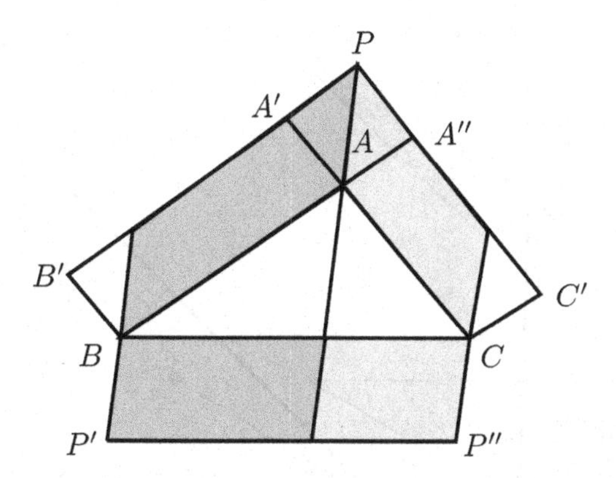

Theorem 2.7.2 (Erdős-Mordell theorem). *Let P be an arbitrary point inside or on the boundary of the triangle ABC. If p_a, p_b, p_c are the distances from P to the sides of ABC, of lenghts a, b, c, respectively, then*

$$PA + PB + PC \geq 2\,(p_a + p_b + p_c)\,.$$

Moreover, the equality holds if and only if the triangle ABC is equilateral and P is the circumcenter.

Proof (Kazarinoff). Let us reflect the triangle ABC on the internal bisector BL of angle B. Let A' and C' be the reflections of A and C. The point P is not reflected. Now, let us consider the parallelograms determined by B, P and A', and by B, P and C'.

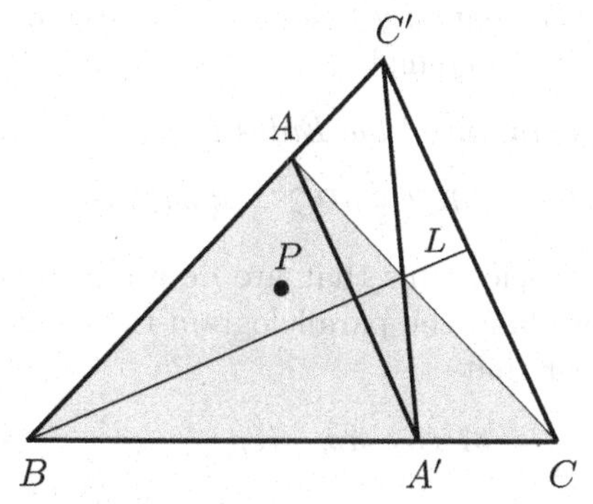

The sum of the areas of these parallelograms is $cp_a + ap_c$ and this is equal to the area of the parallelogram $A'P'P''C'$, where $A'P'$ is parallel to BP and of the same length. The area of $A'P'P''C'$ is at most $b \cdot PB$. Moreover, the areas are equal if BP is perpendicular to $A'C'$ and this happens if and only if P is on BO, where O is the circumcenter of ABC.[14] Then,

$$cp_a + ap_c \leq bPB.$$

[14] BP is perpendicular to $A'C'$ if and only if $\angle PBA' = 90^\circ - \angle A'$, but $\angle A' = \angle A$ and $OBC = 90^\circ - \angle A$, then P should be on BO.

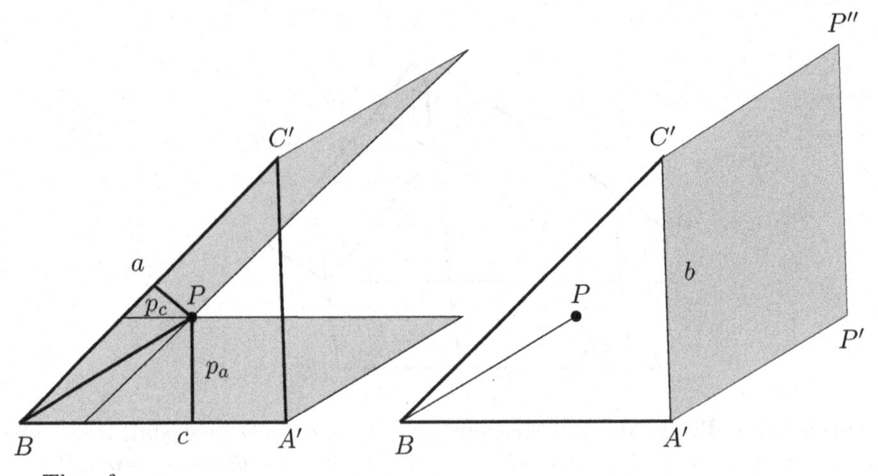

Therefore,

$$PB \geq \frac{c}{b}p_a + \frac{a}{b}p_c.$$

Similarly,

$$PA \geq \frac{b}{a}p_c + \frac{c}{a}p_b \quad \text{and} \quad PC \geq \frac{b}{c}p_a + \frac{a}{c}p_b.$$

If we add together these inequalities, we have

$$PA + PB + PC \geq \left(\frac{b}{c} + \frac{c}{b}\right)p_a + \left(\frac{c}{a} + \frac{a}{c}\right)p_b + \left(\frac{a}{b} + \frac{b}{a}\right)p_c \geq 2\left(p_a + p_b + p_c\right),$$

since $\frac{b}{c} + \frac{c}{b} \geq 2$. Moreover, the equality holds if and only if $a = b = c$ and P is on AO, BO and CO, that is, if the triangle is equilateral and $P = O$. □

Example 2.7.3. *Using the notation of the Erdős-Mordell theorem, prove that*

$$aPA + bPB + cPC \geq 4(ABC).$$

Consider the two parallelograms that are determined by B, C, P and B, A, P as shown in the figure, and the parallelogram that is constructed following Pappus's theorem. It is clear that

$$bPB \geq ap_a + cp_c.$$

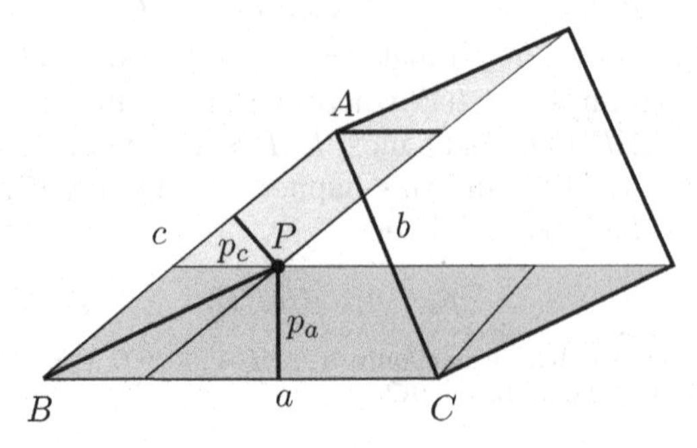

Similarly, it follows that

$$aPA \geq bp_b + cp_c,$$
$$cPC \geq ap_a + bp_b.$$

Hence,

$$aPA + bPB + cPC \geq 2(ap_a + bp_b + cp_c) = 4(ABC).$$

Example 2.7.4. *Using the notation of the Erdős-Mordell theorem, prove that*

$$p_aPA + p_bPB + p_cPC \geq 2\left(p_ap_b + p_bp_c + p_cp_a\right).$$

As in the previous example, we have that $aPA \geq bp_b + cp_c$. Hence,

$$p_aPA \geq \frac{b}{a}p_ap_b + \frac{c}{a}p_cp_a.$$

Similarly, we can deduce that $p_bPB \geq \frac{a}{b}p_ap_b + \frac{c}{b}p_bp_c$, $p_cPC \geq \frac{a}{c}p_cp_a + \frac{b}{c}p_bp_c$.

If we add together these three inequalities, we get

$$\begin{aligned} p_aPA + p_bPB + p_cPC &\geq \left(\frac{a}{b} + \frac{b}{a}\right)p_ap_b + \left(\frac{b}{c} + \frac{c}{b}\right)p_bp_c + \left(\frac{c}{a} + \frac{a}{c}\right)p_cp_a \\ &\geq 2\left(p_ap_b + p_bp_c + p_cp_a\right). \end{aligned}$$

Example 2.7.5. *Using the notation of the Erdős-Mordell theorem, prove that*

$$2\left(\frac{1}{PA} + \frac{1}{PB} + \frac{1}{PC}\right) \leq \frac{1}{p_a} + \frac{1}{p_b} + \frac{1}{p_c}.$$

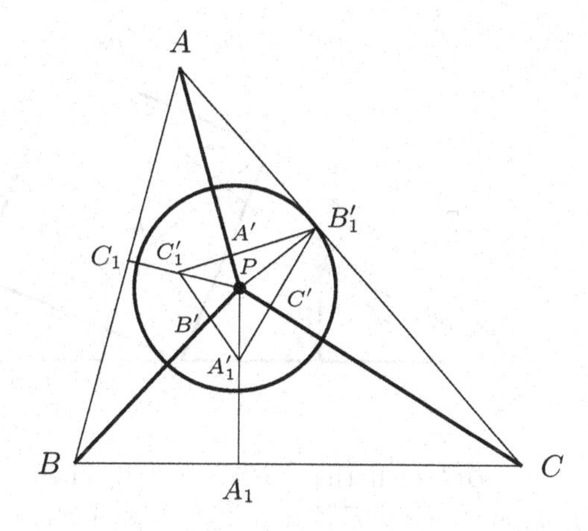

Let us apply inversion to the circle of center P and radius $d = p_b$. If A', B', C' are the inverse points of A, B, C, respectively, and A'_1, B'_1, C'_1 are the inverse points of A_1, B_1, C_1, we can deduce that

$$PA \cdot PA' = PB \cdot PB' = PC \cdot PC' = d^2,$$
$$PA_1 \cdot PA_1' = PB_1 \cdot PB_1' = PC_1 \cdot PC_1' = d^2.$$

Moreover, A', B' and C' are on $B_1'C_1'$, $C_1'A_1'$ and $A_1'B_1'$, respectively, and the segments PA', PB' and PC' are perpendicular to $B_1'C_1'$, $C_1'A_1'$ and $A_1'B_1'$, respectively.

An application of the Erdős-Mordell theorem to the triangle $A_1'B_1'C_1'$ shows that $PA_1' + PB_1' + PC_1' \geq 2\,(PA' + PB' + PC')$.

Since

$$PA_1' = \frac{d^2}{PA_1}, \quad PB_1' = \frac{d^2}{PB_1}, \quad PC_1' = \frac{d^2}{PC_1},$$

$$PC' = \frac{d^2}{PC}, \quad PB' = \frac{d^2}{PB}, \quad PA' = \frac{d^2}{PA},$$

then

$$d^2\left(\frac{1}{PA_1} + \frac{1}{PB_1} + \frac{1}{PC_1}\right) \geq 2d^2\left(\frac{1}{PA} + \frac{1}{PB} + \frac{1}{PC}\right),$$

that is,

$$2\left(\frac{1}{PA} + \frac{1}{PB} + \frac{1}{PC}\right) \leq \left(\frac{1}{p_a} + \frac{1}{p_b} + \frac{1}{p_c}\right).$$

Example 2.7.6. *Using the notation of the Erdős-Mordell theorem, prove that*

$$PA \cdot PB \cdot PC \geq \frac{R}{2r}\,(p_a + p_b)\,(p_b + p_c)\,(p_c + p_a)\,.$$

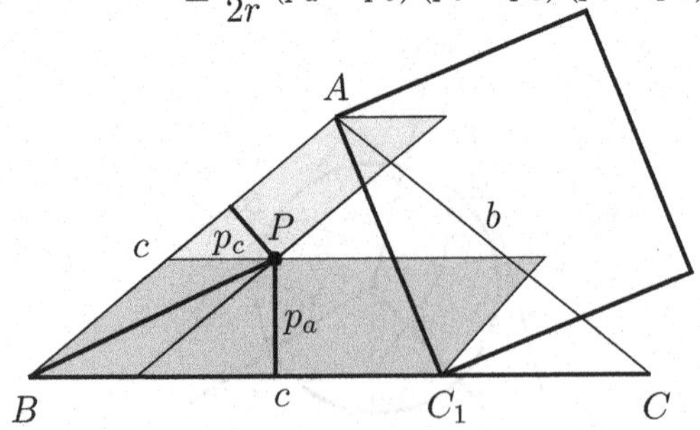

Let C_1 be a point on BC such that $BC_1 = AB$. Then $AC_1 = 2c\sin\frac{B}{2}$, and Pappus's theorem implies that $PB\left(2c\sin\frac{B}{2}\right) \geq c\,p_a + c\,p_c$. Therefore,

$$PB \geq \frac{p_a + p_c}{2\sin\frac{B}{2}}.$$

Similarly,

$$PA \geq \frac{p_b + p_c}{2 \sin \frac{A}{2}} \quad \text{and} \quad PC \geq \frac{p_a + p_b}{2 \sin \frac{C}{2}}.$$

Then, after multiplication, we get

$$PA \cdot PB \cdot PC \geq \frac{1}{8} \frac{1}{\left(\sin \frac{A}{2}\right)\left(\sin \frac{B}{2}\right)\left(\sin \frac{C}{2}\right)} (p_a + p_b)(p_b + p_c)(p_c + p_a).$$

The solution of Exercise 2.42 helps us to prove that $\left(\sin \frac{A}{2}\right)\left(\sin \frac{B}{2}\right)\left(\sin \frac{C}{2}\right) = \frac{r}{4R}$, then the result follows.

Example 2.7.7. (IMO, 1991) *Let P be a point inside the triangle ABC. Prove that at least one of the angles $\angle PAB$, $\angle PBC$, $\angle PCA$ is less than or equal to $30°$.*

Draw A_1, B_1 and C_1, the projections of P on sides BC, CA and AB, respectively. Using the Erdős-Mordell theorem we get $PA + PB + PC \geq 2PA_1 + 2PB_1 + 2PC_1$.

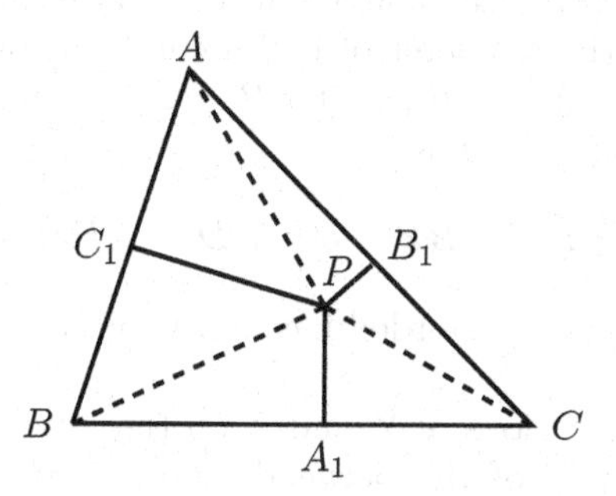

Thus, one of the following inequalities will be satisfied:

$$PA \geq 2PC_1, \quad PB \geq 2PA_1 \quad \text{or} \quad PC \geq 2PB_1.$$

If, for instance, $PA \geq 2PC_1$, we can deduce that $\frac{1}{2} \geq \frac{PC_1}{PA} = \sin \angle PAB$, then $\angle PAB \leq 30°$ or $\angle PAB \geq 150°$. But, if $\angle PAB \geq 150°$, then it must be the case that $\angle PBC < 30°$ and thus in both cases the result follows.

Example 2.7.8. (IMO, 1996) *Let $ABCDEF$ be a convex hexagon such that AB is parallel to DE, BC is parallel to EF and CD is parallel to FA. Let R_A, R_C, R_E denote the circumradii of triangles FAB, BCD, DEF, respectively, and let $\mathcal{P}$ denote the perimeter of the hexagon. Prove that*

$$R_A + R_C + R_E \geq \frac{\mathcal{P}}{2}.$$

Let M, N and P be points inside the hexagon in such a way that $MDEF$, $NFAB$ and $PBCD$ are parallelograms. Let XYZ be the triangle formed by the lines through B, D, F and perpendicular to FA, BC, DE, respectively, where B is on YZ, D on ZX and F on XY. Observe that MNP and XYZ are similar triangles.

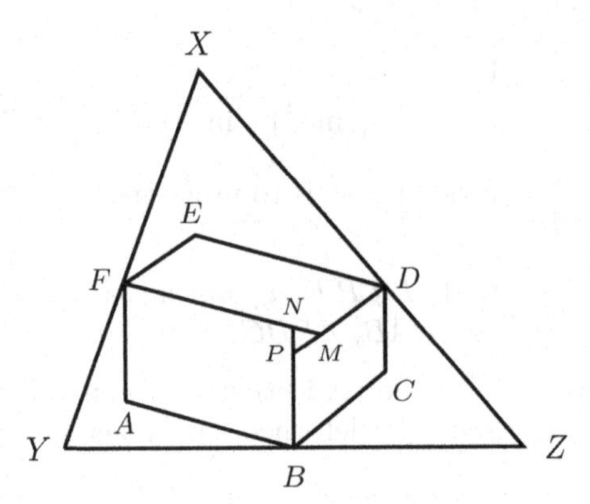

Since the triangles DEF and DMF are congruent, they have the same circumradius; moreover, since XM is the diameter of the circumcircle of triangle DMF, then $XM = 2R_E$. Similarly, $YN = 2R_A$ and $ZP = 2R_C$. Thus, the inequality that needs to be proven can be written as

$$XM + YN + ZP \geq BN + BP + DP + DM + FM + FN.$$

The case $M = N = P$ is the Erdős-Mordell inequality, on which the rest of the proof is based.

Let Y', Z' denote the reflections of Y and Z on the internal angle bisector of X. Let G, H denote the feet of the perpendiculars of M and X on $Y'Z'$, respectively.

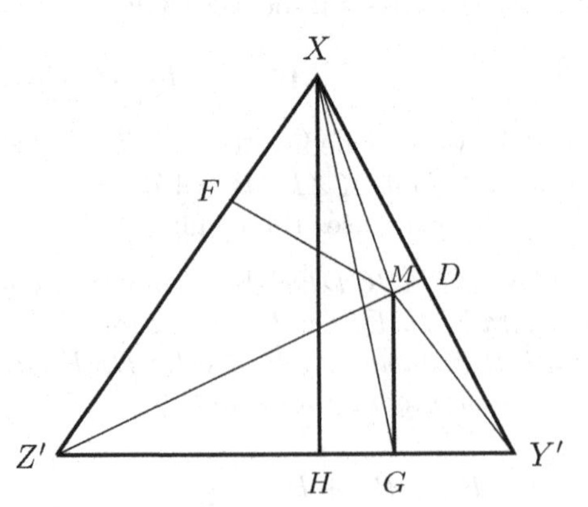

Since $(XY'Z') = (XMZ') + (Z'MY') + (Y'MX)$, we obtain

$$XH \cdot Y'Z' = MF \cdot XZ' + MG \cdot Y'Z' + MD \cdot Y'X.$$

If we set $x = Y'Z'$, $y = ZX'$, $z = XY'$, the above equality becomes

$$xXH = xMG + zDM + yFM.$$

Since $\angle XHG = 90°$, then $XH = XG \sin \angle XGH \leq XG$. Moreover, using the triangle inequality, $XG \leq XM + MG$, we can deduce that

$$XM \geq XH - MG = \frac{z}{x}DM + \frac{y}{x}FM.$$

Similarly,

$$YN \geq \frac{x}{y}FN + \frac{z}{y}BN,$$
$$ZP \geq \frac{y}{z}BP + \frac{x}{y}DP.$$

After adding together these three inequalities, we get

$$XM + YN + ZP \geq \frac{z}{x}DM + \frac{y}{x}FM + \frac{x}{y}FN + \frac{z}{y}BN + \frac{y}{z}BP + \frac{x}{z}DP. \quad (2.9)$$

Observe that

$$\frac{y}{z}BP + \frac{z}{y}BN = \left(\frac{y}{z} + \frac{z}{y}\right)\left(\frac{BP + BN}{2}\right) + \left(\frac{y}{z} - \frac{z}{y}\right)\left(\frac{BP - BN}{2}\right).$$

Since the triangles XYZ and MNP are similar, we can define r as

$$r = \frac{FM - FN}{XY} = \frac{BN - BP}{YZ} = \frac{DP - DM}{ZX}.$$

If we apply the inequality $\frac{y}{z} + \frac{z}{y} \geq 2$, we get

$$\begin{aligned} \frac{y}{z}BP + \frac{z}{y}BN &= \left(\frac{y}{z} + \frac{z}{y}\right)\left(\frac{BP + BN}{2}\right) - \frac{r}{2}\left(\frac{yx}{z} - \frac{zx}{y}\right) \\ &\geq BP + BN - \frac{r}{2}\left(\frac{yx}{z} - \frac{zx}{y}\right). \end{aligned}$$

Similar inequalities hold for

$$\frac{x}{y}FN + \frac{y}{x}FM \geq FN + FM - \frac{r}{2}\left(\frac{xz}{y} - \frac{yz}{x}\right),$$
$$\frac{z}{x}DM + \frac{x}{z}DP \geq DM + DP - \frac{r}{2}\left(\frac{zy}{x} - \frac{xy}{z}\right).$$

If we add the inequalities and substitute them in (2.9), we have

$$XM + YN + ZP \geq BN + BP + DP + DM + FM + FN,$$

which completes the proof.

Exercise 2.70. Using the notation of the Erdős-Mordell theorem, prove that

$$PA \cdot PB \cdot PC \geq \frac{4R}{r} p_a p_b p_c.$$

Exercise 2.71. Using the notation of the Erdős-Mordell theorem, prove that

(i) $\dfrac{PA^2}{p_b p_c} + \dfrac{PB^2}{p_c p_a} + \dfrac{PC^2}{p_a p_b} \geq 12,$

(ii) $\dfrac{PA}{p_b + p_c} + \dfrac{PB}{p_c + p_a} + \dfrac{PC}{p_a + p_b} \geq 3,$

(iii) $\dfrac{PA}{\sqrt{p_b p_c}} + \dfrac{PB}{\sqrt{p_c p_a}} + \dfrac{PC}{\sqrt{p_a p_b}} \geq 6,$

(iv) $PA \cdot PB + PB \cdot PC + PC \cdot PA \geq 4(p_a p_b + p_b p_c + p_c p_a).$

Exercise 2.72. Let ABC be a triangle, P be an arbitrary point in the plane and let p_a, p_b y p_c be the distances from P to the sides of a triangle of lengths a, b and c, respectively. If, for example, P and A are on different sides of the segment BC, then p_a is negative, and we have a similar situation for the other cases. Prove that

$$PA + PB + PC \geq \left(\frac{b}{c} + \frac{c}{b}\right) p_a + \left(\frac{c}{a} + \frac{a}{c}\right) p_b + \left(\frac{a}{b} + \frac{b}{a}\right) p_c.$$

2.8 Optimization problems

In this section we present two classical examples known as the Fermat-Steiner problem and the Fagnano problem.

The Fermat-Steiner problem. This problem seeks to find a point in the interior or on the sides of a triangle such that the sum of the distances from the point to the vertices of the triangle is minimum. We will present three solutions and point out the methods used to solve the problem.

Torricelli's solution. It takes as its starting point the following two lemmas.

Lemma 2.8.1 (Viviani's lemma). *The sum of the distances from an interior point to the sides of an equilateral triangle is equal to the altitude of the triangle.*

Proof. Let P be a point in the interior of the triangle ABC. Draw the triangle $A'B'C'$ with sides parallel to the sides of ABC, with P on $C'A'$ and $B'C'$ on the line through B and C.

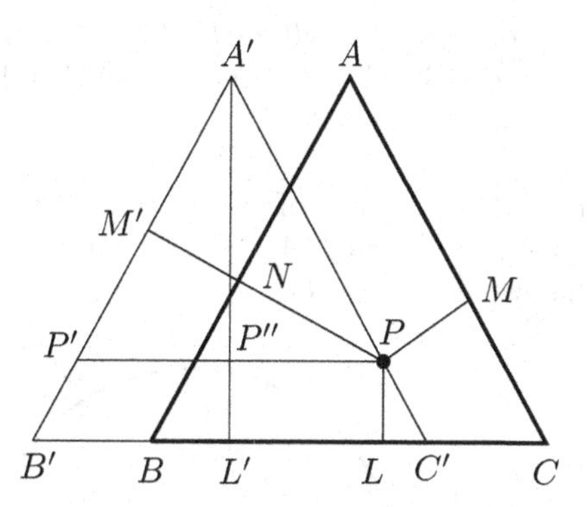

If L, M and N are the feet of the perpendiculars of P on the sides, it is clear that $PM = NM'$, where M' is the intersection of PN with $A'B'$. Moreover, PM' is the altitude of the equilateral triangle $AP'P$. If $A'P''$ is the altitude of the triangle $AP'P$ from A', it is clear that $PM' = A'P''$. Let L' be a point on $B'C'$ such that $A'L'$ is the altitude of the triangle $A'B'C'$ from A'. Thus,

$$PL + PM + PN = PL + PN + NM' = PL + A'P'' = A'P'' + P''L' = A'L'. \quad \square$$

Next, we present another two proofs of Viviani's lemma for the sake of completeness.

Observation 2.8.2. (i) *The following is another proof of Viviani's lemma which is based on the use of areas. We have that* $(ABC) = (ABP) + (BCP) + (CAP)$. *Then, if* a *is the length of the side of the triangle and* h *is the length of its altitude, we have that* $ah = aPN + aPL + aPM$, *that is,* $h = PN + PL + PM$.

(ii) *Another proof of Viviani's lemma can be deduced from the following diagram.*

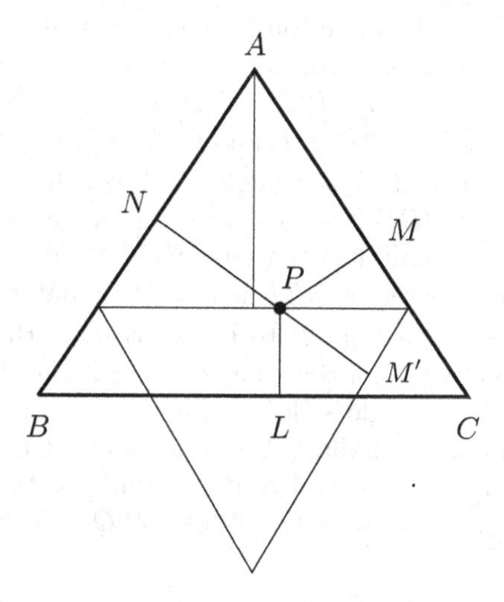

Lemma 2.8.3. *If ABC is a triangle with all angles less than or equal to $120°$, there is a unique point P such that $\angle APB = \angle BPC = \angle CPA = 120°$. The point P is known as the Fermat point of the triangle.*

Proof. First, we will proof the existence of P. On the sides AB and CA we construct equilateral triangles ABC' and CAB'. Their circumcircles intersect at A and at another point that we denote as P.

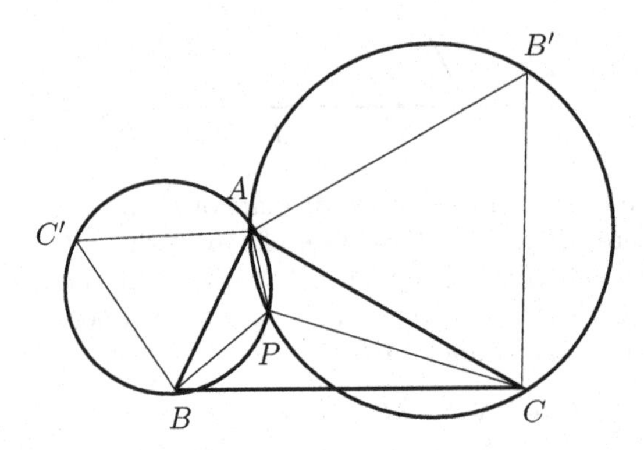

Since $APCB'$ is cyclic, we have that $\angle CPA = 180° - \angle B' = 120°$. Similarly, since $APBC'$ is cyclic, $\angle APB = 120°$. Finally, $\angle BPC = 360° - \angle APB - \angle CPA = 360° - 120° - 120° = 120°$.

To prove the uniqueness, suppose that Q satisfies $\angle AQB = \angle BQC = \angle CQA = 120°$. Since $\angle AQB = 120°$, the point Q should be on the circumcircle of ABC'. Similarly, it should be on the circumcircle of CAB', hence $Q = P$. □

We will now study Torricelli's solution to the Fermat-Steiner problem. Given the triangle ABC with angles less than or equal to $120°$, construct the Fermat point P, which satisfies $\angle APB = \angle BPC = \angle CPA = 120°$. Now, through A, B and C we draw perpendiculars to AP, BP and CP, respectively.

These perpendiculars determine a triangle DEF which is equilateral. This is so because the quadrilateral $PBDC$ is cyclic, having angles of $90°$ in B and C. Now, since $\angle BPC = 120°$, we can deduce that $\angle BDC = 60°$. This argument can be repeated for each angle. Therefore DEF is indeed equilateral.

We know that the distance from P to the vertices of the triangle ABC is equal to the length of the altitude of the equilateral triangle DEF. Observe that any other point Q inside the triangle ABC satisfies $AQ \geq A'Q$, where $A'Q$ is the distance from Q to the side EF, similarly $BQ \geq B'Q$ and $CQ \geq C'Q$. Therefore $AQ + BQ + CQ$ is greater than or equal to the altitude of DEF which is $AP + BP + CP$, which in turn is equal to $A'Q + B'Q + C'Q$ as can be seen by using Viviani's lemma.

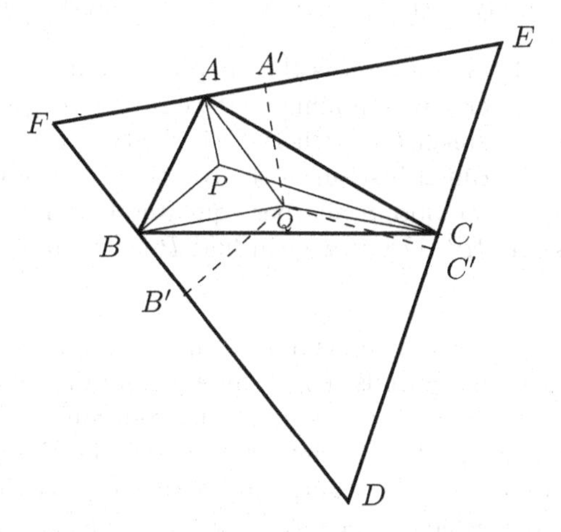

Hofmann-Gallai's solution. This way of solving the problem uses the ingenious idea of rotating the figure to place the three segments that we need next to each other, in order to form a polygonal line and then add them together. Thus, when we join the two extreme points with a segment of line, since this segment of line represents the shortest path between them, it is then necessary to find the conditions under which the polygonal line lies over such segment. This proof was provided by J. Hofmann in 1929, but the method for proving had already been discovered and should be attributed to the Hungarian Tibor Gallai. Let us recall this solution.

Consider the triangle ABC with a point P inside it; draw AP, BP and CP. Next, rotate the figure with its center in B and through an angle of $60°$, in a positive direction.

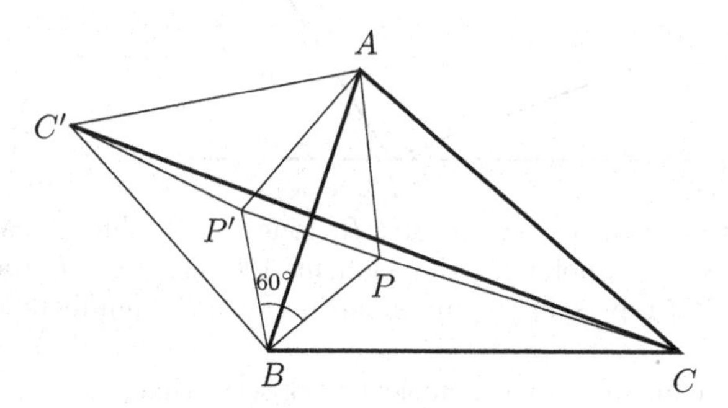

We should point out several things. If C' is the image of A and P' is the image of P, the triangles BPP' and BAC' are equilateral. Moreover, if $AP = P'C'$ and $BP = P'B = P'P$, then $AP + BP + CP = P'C' + P'P + CP$. The path $CP + PP' + P'C$ is minimum when C, P, P' and C' are collinear, which in turn requires that $\angle C'P'B = 120°$ and $\angle BPC = 120°$; but since $\angle C'P'B = \angle APB$, the point P should satisfy $\angle APB = \angle BPC = 120°$ (and then also $\angle CPA = 120°$).

An advantage of this solution is that it provides another description of the Fermat point and another way of finding it. If we review the proof, we can see that the point P is on the segment CC', where C' is the third vertex of the equilateral triangle with side AB. But if instead of the rotation with center in B, we rotate it with its center in C, we obtain another equilateral triangle $AB'C$ and we can conclude that P is on BB'. Hence we can find P as the intersection of BB' and CC'.

Steiner's solution. When we solve maximum and minimum problems we are principally faced with three questions, (i) is there a solution?, (ii) is there a unique solution? (iii) what properties characterize the solution(s)? Torricelli's solution demonstrates that among all the points in the triangle, this particular point P, from which the three sides of the triangle are observed as having an angle of 120°, provides the minimum value of $PA + PB + PC$. In this sense, this point answers the three questions we proposed and does so in an elegant way. However, the solution does not give us any clue as to why Torricelli chose this point, or what made him choose that point; probably this will never be known. But in the following we can consider a sequence of ideas that bring us to discover that the Fermat point is the optimal point. These ideas belong to the Swiss geometer Jacob Steiner. Let us first provide the following two lemmas.

Lemma 2.8.4 (Heron's problem). *Given two points A and B on the same side of a line d, find the shortest path that begins at A, touches the line d and finishes at B.*

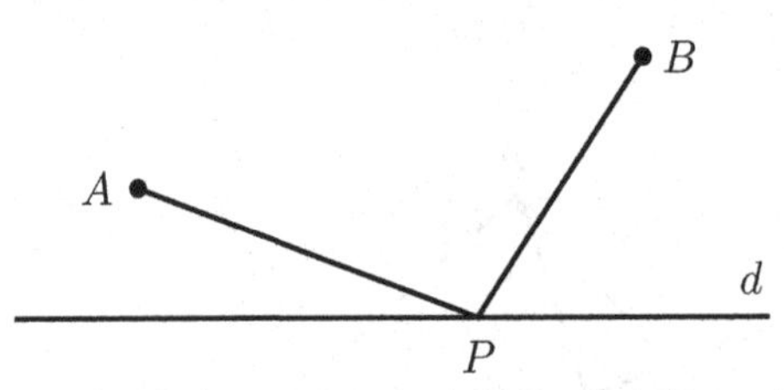

The shortest path between A and B, touching the line d, can be found reflecting B on d to get a point B'; the segment AB' intersects d at a point P^* that makes $AP^* + P^*B$ represent the minimum between the numbers $AP + PB$, with P on d.

To convince ourselves it is sufficient to observe that

$$AP^* + P^*B = AP^* + P^*B' = AB' \leq AP + PB' = AP + PB.$$

This point satisfies the following reflection principle: The incident angle is equal to the reflection angle. It is evident that the point which has this property is the minimum.

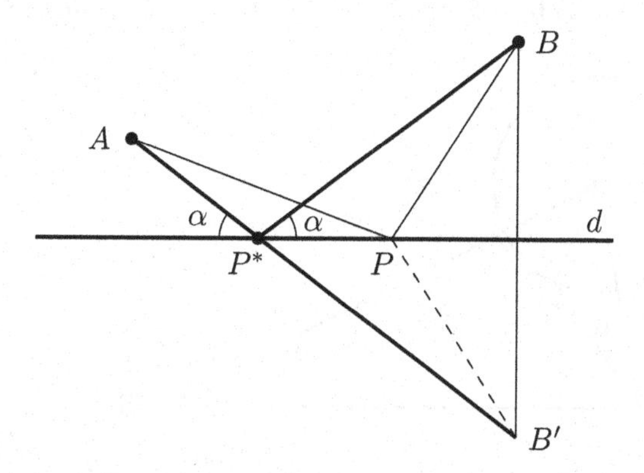

Lemma 2.8.5 (Heron's problem using a circle). *Given two points A and B outside the circle $\mathcal{C}$, find the shortest path that starts at A, touches the circle and finishes at B.*

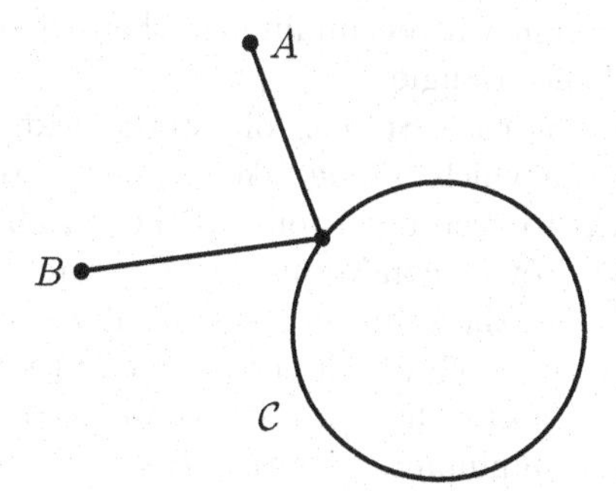

We will only give a sketch of the solution.

Let D be a point on $\mathcal{C}$, then we have that the set $\{P : PA+PB = DA+DB\}$ is an ellipse E_D with foci points A and B, and that the point D belongs to E_D. In general $\mathcal{E}_d = \{P : PA + PB = d\}$, where d is a positive number, is an ellipse with foci A and B (if $d > AB$). Moreover, these ellipses have the property that $\mathcal{E}_d$ is a subset of the interior of $\mathcal{E}_{d'}$ if and only if $d < d'$.

We would like to find a point Q on $\mathcal{C}$ such that the sum $QA+QB$ is minimum. The optimal point Q will belong to an ellipse, precisely to $\mathcal{E}_Q$. Such an ellipse $\mathcal{E}_Q$ does not intersect $\mathcal{C}$ in other point; in fact, if C' is another common point of $\mathcal{E}_Q$ and $\mathcal{C}$, then every point C'' of the circumference arc between Q and C' of $\mathcal{C}$ would be in the interior of $\mathcal{E}_Q$, therefore $C''A + C''B < QA + QB$ and so Q is not the optimal point, that is, a contradiction.

Thus, the point Q that minimize $AQ + QB$ should satisfy that the ellipse $\mathcal{E}_Q$ is tangent to $\mathcal{C}$. The common tangent line to $\mathcal{E}_Q$ and $\mathcal{C}$ happens to be perpendicular to the radius CQ, where C is the center of $\mathcal{C}$ and, because of the reflection property of the ellipse (the incidence angle is equal to the reflection angle), it follows that the line CQ is the internal bisector of the angle $\angle AQB$, that is, $\angle BQC = \angle CQA$.

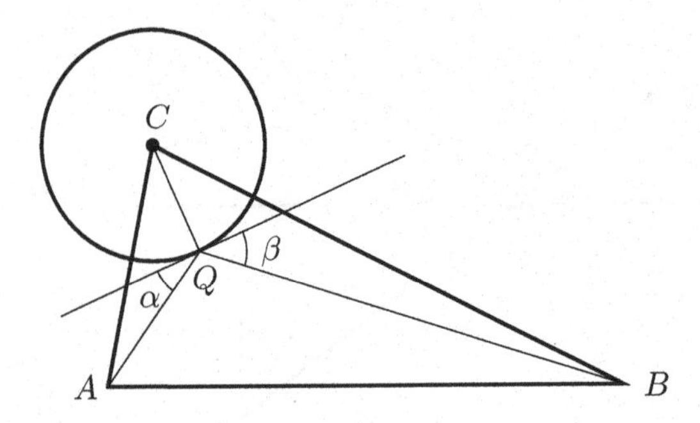

Now let us go back to Steiner's solution of the Fermat-Steiner problem. A point P that makes the sum $PA+PB+PC$ a minimum can be one of the vertices A, B, C or a point of the triangle different from the vertices. In the first case, if P is one of the vertices, then one term of the sum $PA+PB+PC$ is zero and the other two are the lengths of the sides of the triangle ABC that have in common the chosen vertex. Hence, the sum will be minimum when the chosen vertex is opposite to the longest side of the triangle.

In order to analize the second case, Steiner follows the next idea (very useful in optimization problems and one which can be taken to belong to the strategy of *"divide and conquer"*), which is to keep fixed some of the variables and optimize the rest. This procedure would provide conditions in the variables not fixed. Such restrictions will act as restrains in the solution space until we reach the optimal solution. Specifically, we proceed as follows. Suppose that PA is fixed; that is, P belongs to the circle of center A and radius PA, where we need to find the point P that makes the sum $PB+PC$ minimum. Note that B should be located outside of such circle, otherwise $PA \geq AB$ and, using the triangle inequality, $PB+PC > BC$. From this, it follows that $PA+PB = PC > AB+BC$, which means B would be a more suitable point (instead of P). Similarly, C should be outside of such circle. Now, since B and C are points outside the circle $\mathcal{C} = (A, PA)$, the optimal point for the problem of minimizing $PB + PC$ with the condition that P is on the circle $\mathcal{C}$ is, by Lemma 2.8.5, a point Q on the circle $\mathcal{C}$, such that this circle is tangent to the ellipse with foci B and C in Q, and the point Q is such that the angles $\angle AQB$ and $\angle CQA$ are equal. Since the role of A, B, C can be exchanged, if now we fix B (and PB), then the optimal point Q will satisfy the condition $\angle AQB = \angle BQC$ and therefore $\angle AQB = \angle BQC = \angle CQA = 120°$. This means Q should be the Fermat point. All the above work in the second case is to assure that Q is inside of ABC, if the angles of the triangle are not greater than $120°$.

The Fagnano problem. The problem is to find an inscribed triangle of minimum perimeter inside an acute triangle. We present two classical solutions, where the reflection on lines play a central role. One is due to H. Schwarz and the other to L. Fejer.

Schwarz's solution. The German mathematician Hermann Schwarz provided the following solution to this problem for which he took as starting point two observations that we present as lemmas. These lemmas will demonstrate that the inscribed triangle with the minimum perimeter is the triangle formed using the feet of the altitudes of the triangle. Such a triangle is known as the *ortic triangle.*

Lemma 2.8.6. *Let ABC be a triangle, and let D, E and F be the feet of the altitudes on BC, CA and AB as they fall from the vertices A, B and C, respectively. Then the triangles ABC, AEF, DBF and DEC are similar.*

Proof. It is sufficient to see that the first two triangles are similar, since the other cases are proved in a similar way.

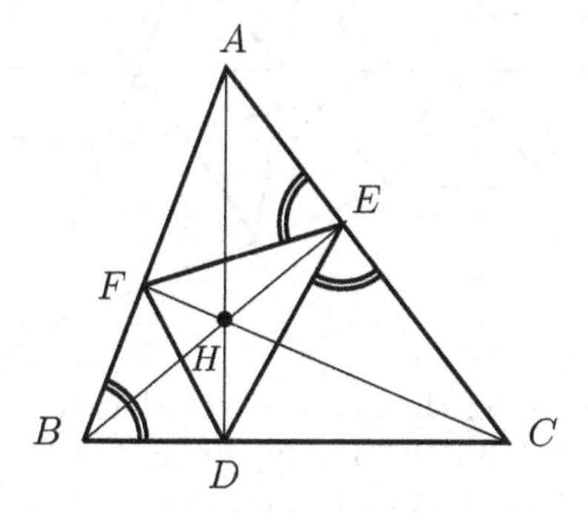

Since these two triangles have a common angle at A, it is sufficient to see that $\angle AEF = \angle ABC$. But, since we know that $\angle AEF + \angle FEC = 180°$ and $\angle ABC + \angle FEC = 180°$ because the quadrilateral $BCEF$ is cyclic, then $\angle AEF = \angle ABC$. □

Lemma 2.8.7. *Using the notation of the previous lemma, we can deduce that the reflection of D on AB is collinear with E and F, and the reflection of D on CA is collinear with E and F.*

Proof. It follows directly from the previous lemma. □

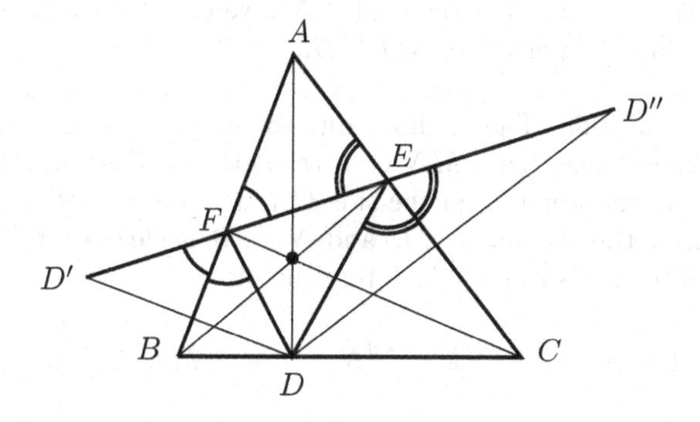

Using these elements we can now continue with the solution proposed by H. Schwarz for the Fagnano problem.

We will now prove that the triangle with minimum perimeter is the ortic triangle. Denote this triangle as DEF and consider another triangle LMN inscribed in ABC.

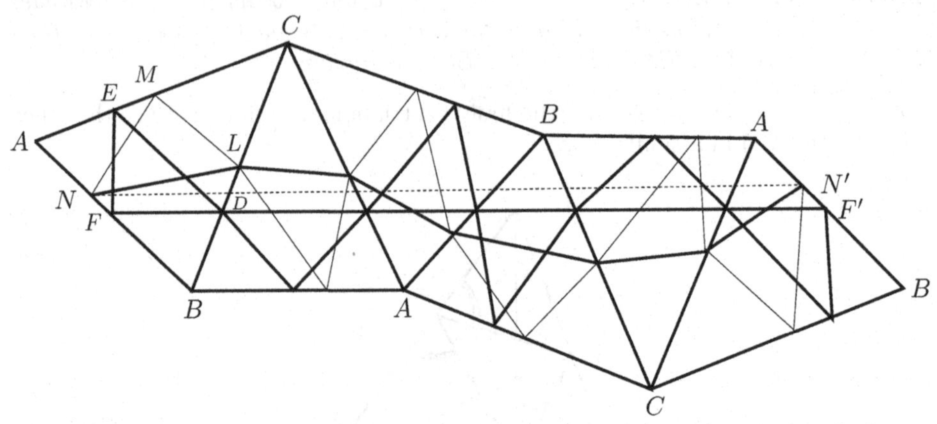

Reflect the complete figure on the side BC, so that the resultant triangle is reflected on CA, then on AB, on BC and finally on CA.

We have in total six congruent triangles and within each of them we have the ortic triangle and the inscribed triangle LMN. The side AB of the last triangle is parallel to the side AB of the first, since as a result of the first reflection, the side AB is rotated in a negative direction through an angle of $2B$, and then in a negative direction through an angle of $2A$, the third reflection is invariant and the fourth is rotated through an angle of $2B$ in a positive direction and in the fifth it is also rotated in a positive direction through an angle of $2A$. Thus the total angle of rotation of AB is zero.

The segment FF' is twice the perimeter of the ortic triangle, since FF' is composed of six pieces where each side of the ortic triangle is taken twice. Also, the broken line NN' is twice the perimeter of LMN. Moreover, NN' is parallel to the line FF' and of the same length, then since the length of the broken line NN' is greater than the length of the segment NN', we can deduce that the perimeter of DEF is less than the perimeter of LMN.

The Fejer's solution. The solution due to the Hungarian mathematician L. Fejer also uses reflections. Let LMN be a triangle inscribed on ABC. Take both the reflection L' of the point L on the side CA, and L'' the reflection of L on the side AB, and draw the segments ML' and NL''. It is clear that $LM = ML'$ and $L''N = NL$. Hence the perimeter of LMN satisfies

$$LM + MN + NL = L''N + NM + ML' \geq L'L''.$$

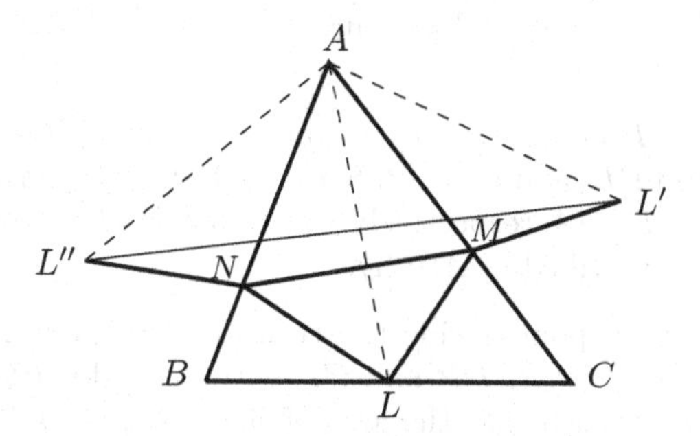

Thus, we can conclude that if we fix the point L, the points M and N that make the minimum perimeter LMN are the intersections of $L'L''$ with CA and AB, respectively. Now, let us see which is the best option for the point L. We already know that the perimeter of LMN is $L'L''$, thus L should make this quantity a minimum.

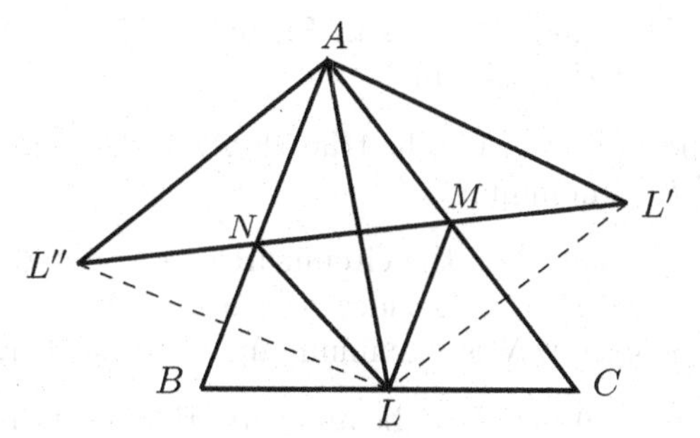

It is evident that $AL = AL' = AL''$ and that AC and AB are internal angle bisectors of the angles LAL' and $L''AL$, respectively. Thus $\angle L''AL' = 2\angle BAC = 2\alpha$ which is a fixed angle. The cosine law applied to the triangle $AL''L'$ guarantees that

$$\begin{aligned}(L'L'')^2 &= (AL')^2 + (AL'')^2 - 2AL' \cdot AL'' \cos 2\alpha \\ &= 2AL^2(1 - \cos 2\alpha).\end{aligned}$$

Then, $L'L''$ is minimum when AL is minimum, which will be the case when AL is the altitude.[15] A similar analysis using the points B and C will demonstrate that

[15]This would be enough to finish Fejer's proof for the Fagnano's problem. This is the case because if AL is the altitude and $L'L''$ intersects sides CA and AB in E and F, respectively, then BE and CF are altitudes. Let us see why. The triangle $AL''L'$ is isosceles with $\angle L''AL' = 2\angle A$, then $\angle AL'L'' = 90^\circ - \angle A$ and by symmetry $\angle ELA = 90^\circ - \angle A$. Therefore $\angle CLE = \angle A$. Then $AELB$ is a cyclic quadrilateral, therefore $\angle AEB = \angle ALB = 90^\circ$, which implies that BE is an altitude. Similarly, it follows that CE is an altitude.

BM and CN should also be altitudes. Thus, the triangle LMN with minimum perimeter is the ortic triangle.

Exercise 2.73. Let $ABCD$ be a convex cyclic quadrilateral. If O is the intersection of the diagonals AC and BD, and P, Q, R, S are the feet of the perpendiculars of O on the sides AB, BC, CD, DA, respectively, prove that $PQRS$ is the quadrilateral of minimum perimeter inscribed in $ABCD$.

Exercise 2.74. Let P be a point inside the triangle ABC. Let D, E and F be the points of intersection of AP, BP and CP with the sides BC, CA and AB, respectively. Determine P such that the area of the triangle DEF is maximum.

Exercise 2.75. (IMO, 1981) Let P be a point inside the triangle ABC. Let D, E, F be the feet of the perpendiculars from P to the lines BC, CA, AB, respectively. Find the point P that minimizes $\frac{BC}{PD} + \frac{CA}{PE} + \frac{AB}{PF}$.

Exercise 2.76. Let P, D, E and F be as in Exercise 2.75. For which point P is the sum of $BD^2 + CE^2 + AF^2$ minimum?

Exercise 2.77. Let P, D, E and F be as in Exercise 2.75. For which point P is the product of $PD \cdot PE \cdot PF$ maximum?

Exercise 2.78. Let P be a point inside the triangle ABC. For which point P is the sum of $PA^2 + PB^2 + PC^2$ minimum?

Exercise 2.79. For every point P on the circumcircle of a triangle ABC, we draw the perpendiculars PM and PN to the sides AB and CA, respectively. Determine for which point P the length MN is maximum and find that length.

Exercise 2.80. (Turkey, 2000) Let ABC be an acute triangle with circumradius R; let h_a, h_b and h_c be the lengths of the altitudes AD, BE and CF, respectively. Let t_a, t_b and t_c be the lengths of the tangents from A, B and C, respectively, to the circumcircle DEF. Prove that

$$\frac{t_a^2}{h_a} + \frac{t_b^2}{h_b} + \frac{t_c^2}{h_c} \leq \frac{3}{2}R.$$

Exercise 2.81. Let h_a, h_b, h_c be the lengths of the altitudes of a triangle ABC, and let p_a, p_b, p_c be the distances from a point P to the sides BC, CA, AB, respectively, where P is a point inside the triangle ABC. Prove that

(i) $\dfrac{h_a}{p_a} + \dfrac{h_b}{p_b} + \dfrac{h_c}{p_c} \geq 9$,

(ii) $h_a h_b h_c \geq 27 p_a p_b p_c$,

(iii) $(h_a - p_a)(h_b - p_b)(h_c - p_c) \geq 8 p_a p_b p_c$.

Exercise 2.82. If h is the length of the largest altitude of an acute triangle, then $r + R \leq h$.

Exercise 2.83. Of all triangles with a common base and the same perimeter, the isosceles triangle has the largest area.

Exercise 2.84. Of all triangles with a given perimeter, the one with largest area is the equilateral triangle.

Exercise 2.85. Of all inscribed triangles on a given circle, the one with largest perimeter is the equilateral triangle.

Exercise 2.86. If P is a point inside the triangle ABC, $l = PA$, $m = PB$ and $n = PC$, prove that

$$(lm + mn + nl)(l + m + n) \geq a^2 l + b^2 m + c^2 n.$$

Exercise 2.87. (IMO, 1961) Let a, b and c be the lengths of the sides of a triangle and let (ABC) be the area of that triangle, prove that

$$4\sqrt{3}(ABC) \leq a^2 + b^2 + c^2.$$

Exercise 2.88. Let (ABC) be the area of a triangle ABC and let F be the Fermat point of the triangle. Prove that

$$4\sqrt{3}(ABC) \leq (AF + BF + CF)^2.$$

Exercise 2.89. Let P be a point inside the triangle ABC, prove that

$$PA + PB + PC \geq 6r.$$

Exercise 2.90. (The area of the pedal triangle). For a triangle ABC and a point P on the plane, we define the *"pedal triangle"* of P with respect to ABC as the triangle $A_1B_1C_1$ where A_1, B_1, C_1 are the feet of the perpendiculars from P to BC, CA, AB, respectively. Prove that

$$(A_1B_1C_1) = \frac{(R^2 - OP^2)(ABC)}{4R^2},$$

where O is the circumcenter. We can thus conclude that the pedal triangle of maximum area is the medial triangle.

Chapter 3

Recent Inequality Problems

Problem 3.1. (Bulgaria, 1995) Let S_A, S_B and S_C denote the areas of the regular heptagons $A_1A_2A_3A_4A_5A_6A_7$, $B_1B_2B_3B_4B_5B_6B_7$ and $C_1C_2C_3C_4C_5C_6C_7$, respectively. Suppose that $A_1A_2 = B_1B_3 = C_1C_4$, prove that

$$\frac{1}{2} < \frac{S_B + S_C}{S_A} < 2 - \sqrt{2}.$$

Problem 3.2. (Czech and Slovak Republics, 1995) Let $ABCD$ be a tetrahedron such that

$$\angle BAC + \angle CAD + \angle DAB = \angle ABC + \angle CBD + \angle DBA = 180^\circ.$$

Prove that $CD \geq AB$.

Problem 3.3. (Estonia, 1995) Let a, b, c be the lengths of the sides of a triangle and let α, β, γ be the angles opposite to the sides. Prove that if the inradius of the triangle is r, then

$$a \sin \alpha + b \sin \beta + c \sin \gamma \geq 9r.$$

Problem 3.4. (France, 1995) Three circles with the same radius pass through a common point. Let S be the set of points which are interior to at least two of the circles. How should the three circles be placed so that the area of S is minimized?

Problem 3.5. (Germany, 1995) Let ABC be a triangle and let D and E be points on BC and CA, respectively, such that DE passes through the incenter of ABC. If $S = \text{area}(CDE)$ and r is the inradius of ABC, prove that $S \geq 2r^2$.

Problem 3.6. (Ireland, 1995) Let A, X, D be points on a line with X between A and D. Let B be a point such that $\angle ABX = 120^\circ$ and let C be a point between B and X. Prove that $2AD \geq \sqrt{3}\,(AB + BC + CD)$.

Problem 3.7. (Korea, 1995) A finite number of points on the plane have the property that any three of them form a triangle with area at most 1. Prove that all these points lie within the interior or on the sides of a triangle with area less than or equal to 4.

Problem 3.8. (Poland, 1995) For a fixed positive integer n, find the minimum value of the sum

$$x_1 + \frac{x_2^2}{2} + \frac{x_3^3}{3} + \cdots + \frac{x_n^n}{n},$$

given that $x_1, x_2, \ldots, x_n$ are positive numbers satisfying that the sum of their reciprocals is n.

Problem 3.9. (IMO, 1995) Let $ABCDEF$ be a convex hexagon with $AB = BC = CD$ and $DE = EF = FA$ such that $\angle BCD = \angle EFA = \frac{\pi}{3}$. Let G and H be points in the interior of the hexagon such that $\angle AGB = \angle DHE = \frac{2\pi}{3}$. Prove that

$$AG + GB + GH + DH + HE \geq CF.$$

Problem 3.10. (Balkan, 1996) Let O be the circumcenter and G the centroid of the triangle ABC. Let R and r be the circumradius and the inradius of the triangle. Prove that $OG \leq \sqrt{R(R-2r)}$.

Problem 3.11. (China, 1996) Suppose that $x_0 = 0$, $x_i > 0$ for $i = 1, 2, \ldots, n$, and $\sum_{i=1}^n x_i = 1$. Prove that

$$1 \leq \sum_{i=1}^{n} \frac{x_i}{\sqrt{1 + x_0 + \cdots + x_{i-1}}\sqrt{x_i + \cdots + x_n}} < \frac{\pi}{2}.$$

Problem 3.12. (Poland, 1996) Let $n \geq 2$ and $a_1, a_2, \ldots, a_n \in \mathbb{R}^+$ with $\sum_{i=1}^n a_i = 1$. Prove that for $x_1, x_2, \ldots, x_n \in \mathbb{R}^+$, with $\sum_{i=1}^n x_i = 1$, we have

$$2\sum_{i<j} x_i x_j \leq \frac{n-2}{n-1} + \sum_{i=1}^{n} \frac{a_i x_i^2}{1 - a_i}.$$

Problem 3.13. (Romania, 1996) Let $x_1, x_2, \ldots, x_n, x_{n+1}$ be positive real numbers with $x_1 + x_2 + \cdots + x_n = x_{n+1}$. Prove that

$$\sum_{i=1}^{n} \sqrt{x_i(x_{n+1} - x_i)} \leq \sqrt{\sum_{i=1}^{n} x_{n+1}(x_{n+1} - x_i)}.$$

Problem 3.14. (St. Petersburg, 1996) Let M be the intersection of the diagonals of a cyclic quadrilateral, let N be the intersection of the segments that join the opposite midpoints and let O be the circumcenter. Prove that $OM \geq ON$.

Problem 3.15. (Austria–Poland, 1996) If w, x, y and z are real numbers satisfying $w + x + y + z = 0$ and $w^2 + x^2 + y^2 + z^2 = 1$, prove that

$$-1 \le wx + xy + yz + zw \le 0.$$

Problem 3.16. (Taiwan, 1997) Let $a_1, \ldots, a_n$ be positive numbers such that $\frac{a_{i-1}+a_{i+1}}{a_i}$ is an integer for all $i = 1, \ldots, n$, $a_0 = a_n$, $a_{n+1} = a_1$ and $n \ge 3$. Prove that

$$2n \le \frac{a_n + a_2}{a_1} + \frac{a_1 + a_3}{a_2} + \frac{a_2 + a_4}{a_3} + \cdots + \frac{a_{n-1} + a_1}{a_n} \le 3n.$$

Problem 3.17. (Taiwan, 1997) Let ABC be an acute triangle with circumcenter O and circumradius R. Prove that if AO intersects the circumcircle of OBC at D, BO intersects the circumcircle of OCA at E and CO intersects the circumcircle of OAB at F, then $OD \cdot OE \cdot OF \ge 8R^3$.

Problem 3.18. (APMO, 1997) Let ABC be a triangle. The internal bisector of the angle in A meets the segment BC at X and the circumcircle at Y. Let $l_a = \frac{AX}{AY}$. Define l_b and l_c in the same way. Prove that

$$\frac{l_a}{\sin^2 A} + \frac{l_b}{\sin^2 B} + \frac{l_c}{\sin^2 C} \ge 3$$

with equality if and only if the triangle is equilateral.

Problem 3.19. (IMO, 1997) Let $x_1, \ldots, x_n$ be real numbers satisfying $|x_1 + \cdots + x_n| = 1$ and $|x_i| \le \frac{n+1}{2}$ for all $i = 1, \ldots, n$. Prove that there exists a permutation $y_1, \ldots, y_n$ of $x_1, \ldots, x_n$ such that

$$|y_1 + 2y_2 + \cdots + ny_n| \le \frac{n+1}{2}.$$

Problem 3.20. (Czech and Slovak Republics, 1998) Let a, b, c be positive real numbers. A triangle exists with sides of lengths a, b and c if and only if there exist numbers x, y and z such that

$$\frac{y}{z} + \frac{z}{y} = \frac{a}{x}, \quad \frac{z}{x} + \frac{x}{z} = \frac{b}{y}, \quad \frac{x}{y} + \frac{y}{x} = \frac{c}{z}.$$

Problem 3.21. (Hungary, 1998) Let $ABCDEF$ be a centrally symmetric hexagon and let P, Q, R be points on the sides AB, CD, EF, respectively. Prove that the area of the triangle PQR is at most one-half of the area of the hexagon.

Problem 3.22. (Iran, 1998) Let x_1, x_2, x_3 and x_4 be positive real numbers such that $x_1x_2x_3x_4 = 1$. Prove that

$$x_1^3 + x_2^3 + x_3^3 + x_4^3 \ge \max\left\{x_1 + x_2 + x_3 + x_4, \frac{1}{x_1} + \frac{1}{x_2} + \frac{1}{x_3} + \frac{1}{x_4}\right\}.$$

Problem 3.23. (Iran, 1998) Let x, y, z be numbers greater than 1 and such that $\frac{1}{x}+\frac{1}{y}+\frac{1}{z}=2$. Prove that

$$\sqrt{x+y+z} \geq \sqrt{x-1}+\sqrt{y-1}+\sqrt{z-1}.$$

Problem 3.24. (Mediterranean, 1998) Let $ABCD$ be a square inscribed in a circle. If M is a point on the arc AB, prove that

$$MC \cdot MD \geq 3\sqrt{3} MA \cdot MB.$$

Problem 3.25. (Nordic, 1998) Let P be a point inside an equilateral triangle ABC of length side a. If the lines AP, BP and CP intersect the sides BC, CA and AB of the triangle at L, M and N, respectively, prove that

$$PL+PM+PN < a.$$

Problem 3.26. (Spain, 1998) A line that contains the centroid G of the triangle ABC intersects the side AB at P and the side CA at Q. Prove that

$$\frac{PB}{PA} \cdot \frac{QC}{QA} \leq \frac{1}{4}.$$

Problem 3.27. (Armenia, 1999) Let O be the center of the circumcircle of the acute triangle ABC. The lines CO, AO and BO intersect the circumcircles of the triangles AOB, BOC and AOC, for the second time, at C_1, A_1 and B_1, respectively. Prove that

$$\frac{AA_1}{OA_1}+\frac{BB_1}{OB_1}+\frac{CC_1}{OC_1} \geq \frac{9}{2}.$$

Problem 3.28. (Balkan, 1999) Let ABC be an acute triangle and let L, M, N be the feet of the perpendiculars from the centroid G of ABC to the sides BC, CA, AB, respectively. Prove that

$$\frac{4}{27} < \frac{(LMN)}{(ABC)} \leq \frac{1}{4}.$$

Problem 3.29. (Belarus, 1999) Let a, b, c be positive real numbers such that $a^2+b^2+c^2=3$. Prove that

$$\frac{1}{1+ab}+\frac{1}{1+bc}+\frac{1}{1+ca} \geq \frac{3}{2}.$$

Problem 3.30. (Czech and Slovak Republics, 1999) For arbitrary positive numbers a, b and c, prove that

$$\frac{a}{b+2c}+\frac{b}{c+2a}+\frac{c}{a+2b} \geq 1.$$

Problem 3.31. (Ireland, 1999) Let a, b, c, d be positive real numbers with $a+b+c+d=1$. Prove that

$$\frac{a^2}{a+b}+\frac{b^2}{b+c}+\frac{c^2}{c+d}+\frac{d^2}{d+a}\geq\frac{1}{2}.$$

Problem 3.32. (Italy, 1999) Let D and E be given points on the sides AB and CA of the triangle ABC such that DE is parallel to BC and DE is tangent to the incircle of ABC. Prove that

$$DE\leq\frac{AB+BC+CA}{8}.$$

Problem 3.33. (Poland, 1999) Let D be a point on the side BC of the triangle ABC such that $AD>BC$. The point E on CA is defined by the equation $\frac{AE}{EC}=\frac{BD}{AD-BC}$. Prove that $AD>BE$.

Problem 3.34. (Romania, 1999) Let a, b, c be positive real numbers such that $ab+bc+ca\leq 3abc$. Prove that $a+b+c\leq a^3+b^3+c^3$.

Problem 3.35. (Romania, 1999) Let x_1, x_2, $\ldots$, x_n be positive real numbers such that $x_1x_2\cdots x_n=1$. Prove that

$$\frac{1}{n-1+x_1}+\frac{1}{n-1+x_2}+\cdots+\frac{1}{n-1+x_n}\leq 1.$$

Problem 3.36. (Romania, 1999) Let $n\geq 2$ be a positive integer and x_1, y_1, x_2, y_2, $\ldots$, x_n, y_n be positive real numbers such that $x_1+x_2+\cdots+x_n\geq x_1y_1+x_2y_2+\cdots+x_ny_n$. Prove that

$$x_1+x_2+\cdots+x_n\leq\frac{x_1}{y_1}+\frac{x_2}{y_2}+\cdots+\frac{x_n}{y_n}.$$

Problem 3.37. (Russia, 1999) Let a, b and c be positive real numbers with $abc=1$. Prove that if $a+b+c\leq\frac{1}{a}+\frac{1}{b}+\frac{1}{c}$, then $a^n+b^n+c^n\leq\frac{1}{a^n}+\frac{1}{b^n}+\frac{1}{c^n}$ for every positive integer n.

Problem 3.38. (Russia, 1999) Let $\{x\}=x-[x]$ denote the fractional part of x. Prove that for every natural number n,

$$\sum_{j=1}^{n^2}\left\{\sqrt{j}\right\}\leq\frac{n^2-1}{2}.$$

Problem 3.39. (Russia, 1999) The positive real numbers x and y satisfy $x^2+y^3\geq x^3+y^4$. Prove that

$$x^3+y^3\leq 2.$$

Problem 3.40. (St. Petersburg, 1999) Let $x_0 > x_1 > \cdots > x_n$ be real numbers. Prove that

$$x_0 + \frac{1}{x_0 - x_1} + \frac{1}{x_1 - x_2} + \cdots + \frac{1}{x_{n-1} - x_n} \geq x_n + 2n.$$

Problem 3.41. (Turkey, 1999) Prove that $(a+3b)(b+4c)(c+2a) \geq 60abc$ for all real numbers $0 \leq a \leq b \leq c$.

Problem 3.42. (United Kingdom, 1999) Three non-negative real numbers a, b and c satisfy $a+b+c=1$. Prove that

$$7(ab+bc+ca) \leq 2 + 9abc.$$

Problem 3.43. (USA, 1999) Let $ABCD$ be a convex cyclic quadrilateral. Prove that

$$|AB - CD| + |AD - BC| \geq 2\,|AC - BD|\,.$$

Problem 3.44. (APMO, 1999) Let $\{a_n\}$ be a sequence of real numbers satisfying $a_{i+j} \leq a_i + a_j$ for all $i, j = 1, 2, \ldots$. Prove that

$$a_1 + \frac{a_2}{2} + \cdots + \frac{a_n}{n} \geq a_n \quad \text{for all } n \in \mathbb{N}.$$

Problem 3.45. (IMO, 1999) Let $n \geq 2$ be a fixed integer.

(a) Determine the smallest constant C such that

$$\sum_{1 \leq i < j \leq n} x_i x_j (x_i^2 + x_j^2) \leq C \left(\sum_{1 \leq i \leq n} x_i \right)^4$$

for all nonnegative real numbers $x_1, \ldots, x_n$.

(b) For this constant C determine when the equality occurs.

Problem 3.46. (Czech and Slovak Republics, 2000) Prove that for all positive real numbers a and b,

$$\sqrt[3]{\frac{a}{b}} + \sqrt[3]{\frac{b}{a}} \leq \sqrt[3]{2(a+b)\left(\frac{1}{a} + \frac{1}{b}\right)}.$$

Problem 3.47. (Korea, 2000) The real numbers a, b, c, x, y, z satisfy $a \geq b \geq c > 0$ and $x \geq y \geq z > 0$. Prove that

$$\frac{a^2x^2}{(by+cz)(bz+cy)} + \frac{b^2y^2}{(cz+ax)(cx+az)} + \frac{c^2z^2}{(ax+by)(ay+bx)} \geq \frac{3}{4}.$$

Problem 3.48. (Mediterranean, 2000) Let P, Q, R, S be the midpoints of the sides BC, CD, DA, AB, respectively, of the convex quadrilateral $ABCD$. Prove that

$$4(AP^2 + BQ^2 + CR^2 + DS^2) \leq 5(AB^2 + BC^2 + CD^2 + DA^2).$$

Problem 3.49. (Austria–Poland, 2000) Let x, y, z be non-negative real numbers such that $x+y+z=1$. Prove that

$$2 \leq (1-x^2)^2 + (1-y^2)^2 + (1-z^2)^2 \leq (1+x)(1+y)(1+z).$$

Problem 3.50. (IMO, 2000) Let a, b, c be positive real numbers with $abc = 1$. Prove that

$$\left(a-1+\frac{1}{b}\right)\left(b-1+\frac{1}{c}\right)\left(c-1+\frac{1}{a}\right) \leq 1.$$

Problem 3.51. (Balkan, 2001) Let a, b, c be positive real numbers such that $abc \leq a+b+c$. Prove that

$$a^2+b^2+c^2 \geq \sqrt{3}\,abc.$$

Problem 3.52. (Brazil, 2001) Prove that $(a+b)(a+c) \geq 2\sqrt{abc(a+b+c)}$, for all positive real numbers a, b, c.

Problem 3.53. (Poland, 2001) Prove that the inequality

$$\sum_{i=1}^{n} ix_i \leq \binom{n}{2} + \sum_{i=1}^{n} x_i^i$$

holds for every integer $n \geq 2$ and for all non-negative real numbers $x_1, x_2, \ldots, x_n$.

Problem 3.54. (Austria–Poland, 2001) Prove that

$$2 < \frac{a+b}{c} + \frac{b+c}{a} + \frac{c+a}{b} - \frac{a^3+b^3+c^3}{abc} \leq 3,$$

where a, b, c are the lengths of the sides of a triangle.

Problem 3.55. (IMO, 2001) Prove that for a, b and c positive real numbers we have

$$\frac{a}{\sqrt{a^2+8bc}} + \frac{b}{\sqrt{b^2+8ca}} + \frac{c}{\sqrt{c^2+8ab}} \geq 1.$$

Problem 3.56. (Short list IMO, 2001) Let x_1, x_2, $\ldots$, x_n be real numbers, prove that

$$\frac{x_1}{1+x_1^2} + \frac{x_2}{1+x_1^2+x_2^2} + \cdots + \frac{x_n}{1+x_1^2+\cdots+x_n^2} < \sqrt{n}.$$

Problem 3.57. (Austria, 2002) Let a, b, c be real numbers such that there exist α, β, $\gamma \in \{-1, 1\}$ with $\alpha a + \beta b + \gamma c = 0$. Determine the smallest positive value of $\left(\frac{a^3+b^3+c^3}{abc}\right)^2$.

Problem 3.58. (Balkan, 2002) Prove that

$$\frac{2}{b(a+b)} + \frac{2}{c(b+c)} + \frac{2}{a(c+a)} \geq \frac{27}{(a+b+c)^2}$$

for positive real numbers a, b, c.

Problem 3.59. (Canada, 2002) Prove that for all positive real numbers a, b, c,

$$\frac{a^3}{bc} + \frac{b^3}{ca} + \frac{c^3}{ab} \geq a + b + c,$$

and determine when equality occurs.

Problem 3.60. (Ireland, 2002) Prove that

$$\frac{x}{1-x} + \frac{y}{1-y} + \frac{z}{1-z} \geq \frac{3\sqrt[3]{xyz}}{1-\sqrt[3]{xyz}}$$

for positive real numbers x, y, z less than 1.

Problem 3.61. (Rioplatense, 2002) Let a, b, c be positive real numbers. Prove that

$$\left(\frac{a}{b+c} + \frac{1}{2}\right)\left(\frac{b}{c+a} + \frac{1}{2}\right)\left(\frac{c}{a+b} + \frac{1}{2}\right) \geq 1.$$

Problem 3.62. (Rioplatense, 2002) Let a, b, c be positive real numbers. Prove that

$$\frac{a+b}{c^2} + \frac{b+c}{a^2} + \frac{c+a}{b^2} \geq \frac{9}{a+b+c} + \frac{1}{a} + \frac{1}{b} + \frac{1}{c}.$$

Problem 3.63. (Russia, 2002) Prove that $\sqrt{x} + \sqrt{y} + \sqrt{z} \geq xy + yz + zx$ for x, y, z positive real numbers such that $x + y + z = 3$.

Problem 3.64. (APMO, 2002) Let a, b, c be positive real numbers satisfying $\frac{1}{a} + \frac{1}{b} + \frac{1}{c} = 1$. Prove that

$$\sqrt{a+bc} + \sqrt{b+ca} + \sqrt{c+ab} \geq \sqrt{abc} + \sqrt{a} + \sqrt{b} + \sqrt{c}.$$

Problem 3.65. (Ireland, 2003) The lengths a, b, c of the sides of a triangle are such that $a + b + c = 2$. Prove that

$$1 \leq ab + bc + ca - abc \leq 1 + \frac{1}{27}.$$

Problem 3.66. (Romania, 2003) Prove that in any triangle ABC the following inequality holds:

$$\frac{1}{m_b m_c} + \frac{1}{m_c m_a} + \frac{1}{m_a m_b} \leq \frac{\sqrt{3}}{S},$$

where S is the area of the triangle and m_a, m_b, m_c are the lengths of the medians.

Problem 3.67. (Romania, 2003) Let a, b, c, d be positive real numbers with $abcd = 1$. Prove that

$$\frac{1+ab}{1+a} + \frac{1+bc}{1+b} + \frac{1+cd}{1+c} + \frac{1+da}{1+d} \geq 4.$$

Problem 3.68. (Romania, 2003) In a triangle ABC, let l_a, l_b, l_c be the lengths of the internal angle bisectors, and let s be the semiperimeter. Prove that

$$l_a + l_b + l_c \leq \sqrt{3}s.$$

Problem 3.69. (Russia, 2003) Let a, b, c be positive real numbers with $a + b + c = 1$. Prove that

$$\frac{1}{1-a} + \frac{1}{1-b} + \frac{1}{1-c} \geq \frac{2}{1+a} + \frac{2}{1+b} + \frac{2}{1+c}.$$

Problem 3.70. (APMO, 2003) Prove that

$$(a^n + b^n)^{\frac{1}{n}} + (b^n + c^n)^{\frac{1}{n}} + (c^n + a^n)^{\frac{1}{n}} < 1 + \frac{2^{\frac{1}{n}}}{2},$$

where $n > 1$ is an integer and a, b, c are the side-lengths of a triangle with unit perimeter.

Problem 3.71. (IMO, 2003) Given $n > 2$ and real numbers $x_1 \leq x_2 \leq \cdots \leq x_n$, prove that

$$\left(\sum_{i,j} |x_i - x_j|\right)^2 \leq \frac{2}{3}(n^2 - 1)\sum_{i,j}(x_i - x_j)^2,$$

where equality holds if and only if $x_1, x_2, \ldots, x_n$ form an arithmetic progression.

Problem 3.72. (Short list Iberoamerican, 2004) If the positive numbers $x_1, x_2, \ldots,$ x_n satisfy $x_1 + x_2 + \cdots + x_n = 1$, prove that

$$\frac{x_1}{x_2(x_1 + x_2 + x_3)} + \frac{x_2}{x_3(x_2 + x_3 + x_4)} + \cdots + \frac{x_n}{x_1(x_n + x_1 + x_2)} \geq \frac{n^2}{3}.$$

Problem 3.73. (Czech and Slovak Republics, 2004) Let $P(x) = ax^2 + bx + c$ be a quadratic polynomial with non-negative real coefficients. Prove that for any positive number x,

$$P(x)P\left(\frac{1}{x}\right) \geq (P(1))^2.$$

Problem 3.74. (Croatia, 2004) Prove that the inequality

$$\frac{a^2}{(a+b)(a+c)} + \frac{b^2}{(b+c)(b+a)} + \frac{c^2}{(c+a)(c+b)} \geq \frac{3}{4}$$

holds for all positive real numbers a, b, c.

Problem 3.75. (Estonia, 2004) Let a, b, c be positive real numbers such that $a^2 + b^2 + c^2 = 3$. Prove that

$$\frac{1}{1+2ab} + \frac{1}{1+2bc} + \frac{1}{1+2ca} \geq 1.$$

Problem 3.76. (Iran, 2004) Let x, y, z be real numbers such that $xyz = -1$, prove that

$$x^4 + y^4 + z^4 + 3(x + y + z) \geq \frac{x^2}{y} + \frac{x^2}{z} + \frac{y^2}{x} + \frac{y^2}{z} + \frac{z^2}{x} + \frac{z^2}{y}.$$

Problem 3.77. (Korea, 2004) Let R and r be the circumradius and the inradius of the acute triangle ABC, respectively. Suppose that $\angle A$ is the largest angle of ABC. Let M be the midpoint of BC and let X be the intersection of the tangents to the circumcircle of ABC at B and C. Prove that

$$\frac{r}{R} \geq \frac{AM}{AX}.$$

Problem 3.78. (Moldova, 2004) Prove that for all real numbers a, b, $c \geq 0$, the following inequality holds:

$$a^3 + b^3 + c^3 \geq a^2\sqrt{bc} + b^2\sqrt{ca} + c^2\sqrt{ab}.$$

Problem 3.79. (Ukraine, 2004) Let x, y, z be positive real numbers with $x+y+z = 1$. Prove that

$$\sqrt{xy + z} + \sqrt{yz + x} + \sqrt{zx + y} \geq 1 + \sqrt{xy} + \sqrt{yz} + \sqrt{zx}.$$

Problem 3.80. (Ukraine, 2004) Let a, b, c be positive real numbers such that $abc \geq 1$. Prove that

$$a^3 + b^3 + c^3 \geq ab + bc + ca.$$

Problem 3.81. (Romania, 2004) Find all positive real numbers a, b, c which satisfy the inequalities

$$4(ab + bc + ca) - 1 \geq a^2 + b^2 + c^2 \geq 3(a^3 + b^3 + c^3).$$

Problem 3.82. (Romania, 2004) The real numbers a, b, c satisfy $a^2 + b^2 + c^2 = 3$. Prove the inequality

$$|a| + |b| + |c| - abc \leq 4.$$

Problem 3.83. (Romania, 2004) Consider the triangle ABC and let O be a point in the interior of ABC. The straight lines OA, OB, OC meet the sides of the triangle at A_1, B_1, C_1, respectively. Let R_1, R_2, R_3 be the radii of the circumcircles of the triangles OBC, OCA, OAB, respectively, and let R be the radius of the circumcircle of the triangle ABC. Prove that

$$\frac{OA_1}{AA_1}R_1 + \frac{OB_1}{BB_1}R_2 + \frac{OC_1}{CC_1}R_3 \geq R.$$

Problem 3.84. (Romania, 2004) Let $n \geq 2$ be an integer and let a_1, a_2, $\ldots$, a_n be real numbers. Prove that for any non-empty subset $S \subset \{1, 2, \ldots, n\}$, the following inequality holds:

$$\left(\sum_{i \in S} a_i\right)^2 \leq \sum_{1 \leq i \leq j \leq n} (a_i + \cdots + a_j)^2.$$

Problem 3.85. (APMO, 2004) For any positive real numbers a, b, c, prove that

$$(a^2+2)(b^2+2)(c^2+2) \geq 9(ab+bc+ca).$$

Problem 3.86. (Short list IMO, 2004) Let a, b and c be positive real numbers such that $ab+bc+ca=1$. Prove that

$$3\sqrt[3]{\frac{1}{abc}+6(a+b+c)} \leq \frac{\sqrt[3]{3}}{abc}.$$

Problem 3.87. (IMO, 2004) Let $n \geq 3$ be an integer. Let $t_1, t_2, \ldots, t_n$ be positive real numbers such that

$$n^2+1 > (t_1+t_2+\cdots+t_n)\left(\frac{1}{t_1}+\frac{1}{t_2}+\cdots+\frac{1}{t_n}\right).$$

Prove that t_i, t_j, t_k are the side-lengths of a triangle for all i, j, k with $1 \leq i < j < k \leq n$.

Problem 3.88. (Japan, 2005) Let a, b and c be positive real numbers such that $a+b+c=1$. Prove that

$$a\sqrt[3]{1+b-c}+b\sqrt[3]{1+c-a}+a\sqrt[3]{1+a-b} \leq 1.$$

Problem 3.89. (Russia, 2005) Let $x_1, x_2, \ldots, x_6$ be real numbers such that $x_1^2+x_2^2+\cdots+x_6^2=6$ and $x_1+x_2+\cdots+x_6=0$. Prove that $x_1x_2\cdots x_6 \leq \frac{1}{2}$.

Problem 3.90. (United Kingdom, 2005) Let a, b, c be positive real numbers. Prove that

$$\left(\frac{a}{b}+\frac{b}{c}+\frac{c}{a}\right)^2 \geq (a+b+c)\left(\frac{1}{a}+\frac{1}{b}+\frac{1}{c}\right).$$

Problem 3.91. (APMO, 2005) Let a, b and c be positive real numbers such that $abc=8$. Prove that

$$\frac{a^2}{\sqrt{(1+a^3)(1+b^3)}}+\frac{b^2}{\sqrt{(1+b^3)(1+c^3)}}+\frac{c^2}{\sqrt{(1+c^3)(1+a^3)}} \geq \frac{4}{3}.$$

Problem 3.92. (IMO, 2005) Let x, y, z be positive real numbers such that $xyz \geq 1$. Prove that

$$\frac{x^5-x^2}{x^5+y^2+z^2}+\frac{y^5-y^2}{y^5+z^2+x^2}+\frac{z^5-z^2}{z^5+x^2+y^2} \geq 0.$$

Problem 3.93. (Balkan, 2006) Let a, b, c be positive real numbers, prove that

$$\frac{1}{a(b+1)}+\frac{1}{b(c+1)}+\frac{1}{c(a+1)} \geq \frac{3}{1+abc}.$$

Problem 3.94. (Estonia, 2006) Let O be the circumcenter of the acute triangle ABC and let A', B' and C' be the circumcenter of the triangles BCO, CAO and ABO, respectively. Prove that the area of the triangle ABC is less than or equal to the area of the triangle $A'B'C'$.

Problem 3.95. (Lithuania, 2006) Let a, b, c be positive real numbers, prove that

$$\frac{1}{a^2+bc}+\frac{1}{b^2+ca}+\frac{1}{c^2+ab}\leq\frac{1}{2}\left(\frac{1}{ab}+\frac{1}{bc}+\frac{1}{ca}\right).$$

Problem 3.96. (Turkey, 2006) Let a_1, a_2, $\ldots$, a_n be positive real numbers such that

$$a_1+a_2+\cdots+a_n=a_1^2+a_2^2+\cdots+a_n^2=A.$$

Prove that

$$\sum_{i\neq j}\frac{a_i}{a_j}\geq\frac{(n-1)^2A}{A-1}.$$

Problem 3.97. (Iberoamerican, 2006) Consider n real numbers a_1, a_2, $\ldots$, a_n, not necessarily distinct. Let d be the difference between the maximum and the minimum value of the numbers and let $s=\sum_{i<j}|a_i-a_j|$. Prove that

$$(n-1)d\leq s\leq\frac{n^2d}{4},$$

and determine the conditions on the n numbers that ensure the validity of the equalities.

Problem 3.98. (IMO, 2006) Determine the least real number M such that the inequality

$$\left|ab(a^2-b^2)+bc(b^2-c^2)+ca(c^2-a^2)\right|\leq M(a^2+b^2+c^2)^2$$

is satisfied for all real numbers a, b, c.

Problem 3.99. (Bulgaria, 2007) Find all positive integers n such that if a, b, c are non-negative real numbers with $a+b+c=3$, then

$$abc(a^n+b^n+c^n)\leq 3.$$

Problem 3.100. (Bulgaria, 2007) If a, b, c are positive real numbers, prove that

$$\frac{(a+1)(b+1)^2}{3\sqrt[3]{c^2a^2}+1}+\frac{(b+1)(c+1)^2}{3\sqrt[3]{a^2b^2}+1}+\frac{(c+1)(a+1)^2}{3\sqrt[3]{b^2c^2}+1}\geq a+b+c+3.$$

Problem 3.101. (China, 2007) If a, b, c are the lengths of the sides of a triangle with $a+b+c=3$, find the minimum of

$$a^2+b^2+c^2+\frac{4abc}{3}.$$

Problem 3.102. (Greece, 2007) If a, b, c are the lengths of the sides of a triangle, prove that

$$\frac{(a+b-c)^4}{b(b+c-a)} + \frac{(b+c-a)^4}{c(c+a-b)} + \frac{(c+a-b)^4}{a(a+b-c)} \geq ab+bc+ca.$$

Problem 3.103. (Iran, 2007) If a, b, c are three different positive real numbers, prove that

$$\left|\frac{a+b}{a-b} + \frac{b+c}{b-c} + \frac{c+a}{c-a}\right| > 1.$$

Problem 3.104. (Mediterranean, 2007) Let x, y, z be real numbers such that $xy+yz+zx = 1$. Prove that $xz < \frac{1}{2}$. Is it possible to improve the bound $\frac{1}{2}$?

Problem 3.105. (Mediterranean, 2007) Let $x > 1$ be a real number which is not an integer. Prove that

$$\left(\frac{x+\{x\}}{[x]} - \frac{[x]}{x+\{x\}}\right) + \left(\frac{x+[x]}{\{x\}} - \frac{\{x\}}{x+[x]}\right) > \frac{9}{2},$$

where $[x]$ and $\{x\}$ represent the integer part and the fractional part of x, respectively.

Problem 3.106. (Peru, 2007) Let a, b, c be positive real numbers such that $a+b+c \geq \frac{1}{a} + \frac{1}{b} + \frac{1}{c}$. Prove that

$$a+b+c \geq \frac{3}{a+b+c} + \frac{2}{abc}.$$

Problem 3.107. (Romania, 2007) Let a, b, c be positive real numbers such that

$$\frac{1}{a+b+1} + \frac{1}{b+c+1} + \frac{1}{c+a+1} \geq 1.$$

Prove that

$$a+b+c \geq ab+bc+ca.$$

Problem 3.108. (Romania, 2007) Let ABC be an acute triangle with $AB = AC$. For every interior point P of ABC, consider the circle with center A and radius AP; let M and N be the intersections of the sides AB and AC with the circle, respectively. Determine the position of P in such a way that $MN + BP + CP$ is minimum.

Problem 3.109. (Romania, 2007) The points M, N, P on the sides BC, CA, AB, respectively, are such that the triangle MNP is acute. Let x be the length of the shortest altitude in the triangle ABC and let X be the length of the largest altitude in the triangle MNP. Prove that $x \leq 2X$.

Problem 3.110. (APMO, 2007) Let x, y, z be positive real numbers such that $\sqrt{x}+\sqrt{y}+\sqrt{z}=1$. Prove that

$$\frac{x^2+yz}{\sqrt{2x^2(y+z)}}+\frac{y^2+zx}{\sqrt{2y^2(z+x)}}+\frac{z^2+xy}{\sqrt{2z^2(x+y)}}\ge 1.$$

Problem 3.111. (Baltic, 2008) If the positive real numbers a, b, c satisfy $a^2+b^2+c^2=3$, prove that

$$\frac{a^2}{2+b+c^2}+\frac{b^2}{2+c+a^2}+\frac{c^2}{2+a+b^2}\ge\frac{(a+b+c)^2}{12}.$$

Under which circumstances the equality holds?

Problem 3.112. (Canada, 2008) Let a, b, c be positive real numbers for which $a+b+c=1$. Prove that

$$\frac{a-bc}{a+bc}+\frac{b-ca}{b+ca}+\frac{c-ab}{c+ab}\le\frac{3}{2}.$$

Problem 3.113. (Iran, 2008) Find the least real number K such that for any positive real numbers x, y, z, the following inequality holds:

$$x\sqrt{y}+y\sqrt{z}+z\sqrt{x}\le K\sqrt{(x+y)(y+z)(z+x)}.$$

Problem 3.114. (Ireland, 2008) If the positive real numbers a, b, c, d satisfy $a^2+b^2+c^2+d^2=1$, prove that

$$a^2b^2cd+ab^2c^2d+abc^2d^2+a^2bcd^2+a^2bc^2d+ab^2cd^2\le\frac{3}{32}.$$

Problem 3.115. (Ireland, 2008) Let x, y, z be positive real numbers such that $xyz\ge 1$. Prove that

(a) $27\le(1+x+y)^2+(1+y+z)^2+(1+z+x)^2$,

(b) $(1+x+y)^2+(1+y+z)^2+(1+z+x)^2\le 3(x+y+z)^2$.

The equalities hold if and only if $x=y=z=1$.

Problem 3.116. (Romania, 2008) If a, b, c are positive real numbers with $ab+bc+ca=3$, prove that

$$\frac{1}{1+a^2(b+c)}+\frac{1}{1+b^2(c+a)}+\frac{1}{1+c^2(a+b)}\le\frac{1}{abc}.$$

Problem 3.117. (Romania, 2008) Determine the maximum value for the real number k if

$$(a+b+c)\left(\frac{1}{a+b}+\frac{1}{b+c}+\frac{1}{c+a}-k\right)\ge k$$

for all real numbers a, b, $c\ge 0$ and with $a+b+c=ab+bc+ca$.

Problem 3.118. (Serbia, 2008) Let a, b, c be positive real numbers such that $a+b+c=1$. Prove that

$$a^2+b^2+c^2+3abc \geq \frac{4}{9}.$$

Problem 3.119. (Vietnam, 2008) Let x, y, z be distinct non-negative real numbers. Prove that

$$\frac{1}{(x-y)^2}+\frac{1}{(y-z)^2}+\frac{1}{(z-x)^2} \geq \frac{4}{xy+yz+zx}.$$

When is the case that the equality holds?

Problem 3.120. (IMO, 2008)

(i) If x, y, z are three real numbers different from 1 and such that $xyz=1$, prove that

$$\frac{x^2}{(x-1)^2}+\frac{y^2}{(y-1)^2}+\frac{z^2}{(z-1)^2} \geq 1.$$

(ii) Prove that the equality holds for an infinite number of x, y, z, all of them being rational numbers.

[illegible]

Theorem 3.13. [illegible] Then [illegible] is positive [illegible]

$$[illegible]$$

[illegible] (2008) [illegible]

$$[illegible]$$

[illegible]

[illegible]

[illegible]

$$[illegible]$$

[illegible] and [illegible]

Chapter 4

Solutions to Exercises and Problems

In this chapter we present solutions or hints to the exercises and problems that appear in this book. In Sections 1 and 2 we provide the solutions to the exercises in Chapters 1 and 2, respectively, and in Section 3 the solutions to the problems in Chapter 3. We recommend that the reader should consult this chapter only after having tried to solve the exercises or the problems by himself.

4.1 Solutions to the exercises in Chapter 1

Solution 1.1. It follows from the definition of $a < b$ and Property 1.1.1 for the number $a - b$.

Solution 1.2. (i) If $a < 0$, then $-a > 0$. Also use $(-a)(-b) = ab$.
(ii) $(-a)b > 0$.
(iii) $a < b \Leftrightarrow b - a > 0$, now use property 1.1.2.
(iv) Use property 1.1.2.
(v) If $a < 0$, then $-a > 0$.
(vi) $a\frac{1}{a} = 1 > 0$.
(vii) If $a < 0$, then $-a > 0$.
(viii) Use (vi) and property 1.1.3.
(ix) Prove that $ac < bc$ and that $bc < bd$.
(x) Use property 1.1.3 with $a - 1 > 0$ and $a > 0$.
(xi) Use property 1.1.3 with $1 - a > 0$ and $a > 0$.

Solution 1.3. (i) $a^2 < b^2 \Leftrightarrow b^2 - a^2 = (b + a)(b - a) > 0$.
(ii) If $b > 0$, then $\frac{1}{b} > 0$, now use Example 1.1.4.

Solution 1.4. For (i), (ii) and (iii) use the definition, and for (iv) and (v) remember that $|a|^2 = a^2$.

Solution 1.5. (i) $x \leq |x|$ and $-x \leq |x|$.
(ii) Consider $|a| = |a - b + b|$ and $|b| = |b - a + a|$, and apply the triangle inequality.
(iii) $(x^2 + xy + y^2)(x - y) = x^3 - y^3$.
(iv) $(x^2 - xy + y^2)(x + y) = x^3 + y^3$.

Solution 1.6. If a, b or c is zero, the equality follows. Then, we can assume $|a| \geq |b| \geq |c| > 0$. Dividing by $|a|$, the inequality is equivalent to

$$1 + \left|\frac{b}{a}\right| + \left|\frac{c}{a}\right| - \left|1 + \frac{b}{a}\right| - \left|\frac{b}{a} + \frac{c}{a}\right| - \left|1 + \frac{c}{a}\right| + \left|1 + \frac{b}{a} + \frac{c}{a}\right| \geq 0.$$

Since $\left|\frac{b}{a}\right| \leq 1$ and $\left|\frac{c}{a}\right| \leq 1$, we can deduce that $\left|1 + \frac{b}{a}\right| = 1 + \frac{b}{a}$ and $\left|1 + \frac{c}{a}\right| = 1 + \frac{c}{a}$. Thus, it is sufficient to prove that

$$\left|\frac{b}{a}\right| + \left|\frac{c}{a}\right| - \left|\frac{b}{a} + \frac{c}{a}\right| - \left(1 + \frac{b}{a} + \frac{c}{a}\right) + \left|1 + \frac{b}{a} + \frac{c}{a}\right| \geq 0.$$

Now, use the triangle inequality and Exercise 1.5.

Solution 1.7. (i) Use that $0 \leq b \leq 1$ and $1 + a > 0$ in order to see that

$$0 \leq b(1 + a) \leq 1 + a \Rightarrow 0 \leq b - a \leq 1 - ab \Rightarrow 0 \leq \frac{b - a}{1 - ab} \leq 1.$$

(ii) The inequality on the left-hand side is clear. Since $1 + a \leq 1 + b$, it follows that $\frac{1}{1+b} \leq \frac{1}{1+a}$, and then prove that

$$\frac{a}{1 + b} + \frac{b}{1 + a} \leq \frac{a}{1 + a} + \frac{b}{1 + a} = \frac{a + b}{1 + a} \leq 1.$$

(iii) For the inequality on the left-hand side, use that $ab^2 - ba^2 = ab(b - a)$ is the product of non-negative real numbers. For the inequality on the right-hand side, note that $b \leq 1 \Rightarrow b^2 \leq b \Rightarrow -b \leq -b^2$, and then

$$ab^2 - ba^2 \leq ab^2 - b^2a^2 = b^2(a - a^2) \leq a - a^2 = \frac{1}{4} - (\frac{1}{2} - a)^2 \leq \frac{1}{4}.$$

Solution 1.8. Prove in general that $x < \sqrt{2} \Rightarrow 1 + \frac{1}{1+x} > \sqrt{2}$ and that $x > \sqrt{2} \Rightarrow 1 + \frac{1}{1+x} < \sqrt{2}$.

Solution 1.9. $ax + by \geq ay + bx \Leftrightarrow (a - b)(x - y) \geq 0$.

Solution 1.10. We can assume that $x \geq y$. Then, use the previous exercise substituting with $\sqrt{x^2}$, $\sqrt{y^2}$, $\frac{1}{\sqrt{y}}$ and $\frac{1}{\sqrt{x}}$.

Solution 1.11. Observe that

$$(a - b)(c - d) + (a - c)(b - d) + (d - a)(b - c) = 2(a - b)(c - d) = 2(a - b)^2 \geq 0.$$

Solution 1.12. It follows from

$$\begin{aligned} f(a,c,b,d)-f(a,b,c,d) &= (a-c)^2-(a-b)^2+(b-d)^2-(c-d)^2 \\ &= (b-c)(2a-b-c)+(b-c)(b+c-2d) \\ &= 2(b-c)(a-d)>0, \end{aligned}$$

$$\begin{aligned} f(a,b,c,d)-f(a,b,d,c) &= (b-c)^2-(b-d)^2+(d-a)^2-(c-a)^2 \\ &= (d-c)(2b-c-d)+(d-c)(c+d-2a) \\ &= 2(d-c)(b-a)>0. \end{aligned}$$

Solution 1.13. In order for the expressions in the inequality to be well defined, it is necessary that $x \geq -\frac{1}{2}$ and $x \neq 0$. Multiply the numerator and the denominator by $(1+\sqrt{1+2x})^2$. Perform some simplifications and show that $2\sqrt{2x+1} < 7$; then solve for x.

Solution 1.14. Since $4n^2 < 4n^2+n < 4n^2+4n+1$, we can deduce that $2n < \sqrt{4n^2+n} < 2n+1$. Hence, its integer part is $2n$ and then we have to prove that $\sqrt{4n^2+n} < 2n+\frac{1}{4}$, this follows immediately after squaring both sides of the inequality.

Solution 1.15. Since $(a^3-b^3)(a^2-b^2) \geq 0$, we have that $a^5+b^5 \geq a^2b^2(a+b)$, then

$$\frac{ab}{a^5+b^5+ab} \leq \frac{ab}{a^2b^2(a+b)+ab} = \frac{abc^2}{a^2b^2c^2(a+b)+abc^2} = \frac{c}{a+b+c}.$$

Similarly, $\frac{bc}{b^5+c^5+bc} \leq \frac{a}{a+b+c}$ and $\frac{ca}{c^5+a^5+ca} \leq \frac{b}{a+b+c}$. Hence,

$$\frac{ab}{a^5+b^5+ab}+\frac{bc}{b^5+c^5+bc}+\frac{ca}{c^5+a^5+ca} \leq \frac{c}{a+b+c}+\frac{a}{a+b+c}+\frac{b}{a+b+c},$$

but $\frac{c}{a+b+c}+\frac{a}{a+b+c}+\frac{b}{a+b+c}=\frac{c+a+b}{a+b+c}=1$.

Solution 1.16. Consider $p(x)=ax^2+bx+c$, using the hypothesis, $p(1)=a+b+c$ and $p(-1)=a-b+c$ are not negative. Since $a>0$, the minimum value of p is attained at $\frac{-b}{2a}$ and its value is $\frac{4ac-b^2}{4a} < 0$. If x_1, x_2 are the roots of p, we can deduce that $\frac{b}{a}=-(x_1+x_2)$ and $\frac{c}{a}=x_1x_2$, therefore $\frac{a+b+c}{a}=(1-x_1)(1-x_2)$, $\frac{a-b+c}{a}=(1+x_1)(1+x_2)$ and $\frac{a-c}{a}=1-x_1x_2$. Observe that, $(1-x_1)(1-x_2) \geq 0$, $(1+x_1)(1+x_2) \geq 0$ and $1-x_1x_2 \geq 0$ imply that $-1 \leq x_1, x_2 \leq 1$.

Solution 1.17. If the inequalities are true, then a, b and c are less than 1, and $a(1-b)b(1-c)c(1-a) > \frac{1}{64}$. On the other hand, since $x(1-x) \leq \frac{1}{4}$ for $0 \leq x \leq 1$, then $a(1-b)b(1-c)c(1-a) \leq \frac{1}{64}$.

Solution 1.18. Use the AM-GM inequality with $a=1$, $b=x$.

Solution 1.19. Use the AM-GM inequality with $a = x$, $b = \frac{1}{x}$.

Solution 1.20. Use the AM-GM inequality with $a = x^2$, $b = y^2$.

Solution 1.21. In the previous exercise add $x^2 + y^2$ to both sides.

Solution 1.22. Use the AM-GM inequality with $a = \frac{x+y}{x}$, $b = \frac{x+y}{y}$ and also use the AM-GM inequality for x and y. Or reduce this to Exercise 1.20.

Solution 1.23. Use the AM-GM inequality with ax and $\frac{b}{x}$.

Solution 1.24. Use the AM-GM inequality with $\frac{a}{b}$ and $\frac{b}{a}$.

Solution 1.25. $\frac{a+b}{2} - \sqrt{ab} = \frac{\left(\sqrt{a}-\sqrt{b}\right)^2}{2}$, simplify and find the bounds using $0 < b \leq a$.

Solution 1.26. $x + y \geq 2\sqrt{xy}$.

Solution 1.27. $x^2 + y^2 \geq 2xy$.

Solution 1.28. $xy + zx \geq 2x\sqrt{yz}$.

Solution 1.29. See Exercise 1.27.

Solution 1.30. $\frac{1}{x} + \frac{1}{y} \geq \frac{2}{\sqrt{xy}}$.

Solution 1.31. $\frac{xy}{z} + \frac{yz}{x} \geq 2\sqrt{\frac{xy^2z}{zx}} = 2y$.

Solution 1.32. $\frac{x^2+(y^2+z^2)}{2} \geq x\sqrt{y^2 + z^2}$.

Solution 1.33. $x^4 + y^4 + 8 = x^4 + y^4 + 4 + 4 \geq 4\sqrt[4]{x^4y^4 16} = 8xy$.

Solution 1.34. $(a + b + c + d) \geq 4\sqrt[4]{abcd}$, $\left(\frac{1}{a} + \frac{1}{b} + \frac{1}{c} + \frac{1}{d}\right) \geq 4\sqrt[4]{\frac{1}{abcd}}$.

Solution 1.35. $\frac{a}{b} + \frac{b}{c} + \frac{c}{d} + \frac{d}{a} \geq 4\sqrt[4]{\frac{a}{b}\frac{b}{c}\frac{c}{d}\frac{d}{a}} = 4$.

Solution 1.36. $(x_1 + \cdots + x_n) \geq n\sqrt[n]{x_1 \cdots x_n}$, $\left(\frac{1}{x_1} + \cdots + \frac{1}{x_n}\right) \geq n\sqrt[n]{\frac{1}{x_1 \cdots x_n}}$.

Solution 1.37. $\frac{a_1}{b_1} + \frac{a_2}{b_2} + \cdots + \frac{a_n}{b_n} \geq n\sqrt[n]{\frac{a_1 \cdots a_n}{b_1 \cdots b_n}} = n$.

Solution 1.38. $a^n - 1 > n\left(a^{\frac{n+1}{2}} - a^{\frac{n-1}{2}}\right) \Leftrightarrow (a-1)\left(a^{n-1} + \cdots + 1\right) > na^{\frac{n-1}{2}}(a-1) \Leftrightarrow \frac{a^{n-1}+\cdots+a+1}{n} > a^{\frac{n-1}{2}}$, but $\frac{1+a+\cdots+a^{n-1}}{n} > \sqrt[n]{a^{\frac{(n-1)n}{2}}} = a^{\frac{n-1}{2}}$.

Solution 1.39. $1 = \left(\frac{1+a}{2}\right)\left(\frac{1+b}{2}\right)\left(\frac{1+c}{2}\right) \geq \sqrt{a}\sqrt{b}\sqrt{c} = \sqrt{abc}$.

Solution 1.40. Using the AM-GM inequality, we obtain

$$\frac{a^3}{b} + \frac{b^3}{c} + bc \geq 3\sqrt[3]{\frac{a^3}{b} \cdot \frac{b^3}{c} \cdot bc} = 3ab.$$

Similarly, $\frac{b^3}{c}+\frac{c^3}{a}+ca \geq 3bc$ and $\frac{c^3}{a}+\frac{a^3}{b}+ab \geq 3ca$. Therefore, $2(\frac{a^3}{b}+\frac{b^3}{c}+\frac{c^3}{a})+(ab+bc+ca) \geq 3(ab+bc+ca)$.

Second solution. The inequality can also be proved using Exercise 1.107.

Solution 1.41. If $abc=0$, the result is clear. If $abc>0$, then we have

$$\begin{aligned}\frac{ab}{c}+\frac{bc}{a}+\frac{ca}{b} &= \frac{1}{2}\left(a\left(\frac{b}{c}+\frac{c}{b}\right)+b\left(\frac{c}{a}+\frac{a}{c}\right)+c\left(\frac{a}{b}+\frac{b}{a}\right)\right)\\ &\geq \frac{1}{2}(2a+2b+2c),\end{aligned}$$

and the result is evident.

Solution 1.42. Apply the AM-GM inequality twice over, $a^2b+b^2c+c^2a \geq 3abc$, $ab^2+bc^2+ca^2 \geq 3abc$.

Solution 1.43. $\frac{1+ab}{1+a}=\frac{abc+ab}{1+a}=ab\left(\frac{1+c}{1+a}\right)$,

$$\begin{aligned}\frac{1+ab}{1+a}+\frac{1+bc}{1+b}+\frac{1+ca}{1+c} &= ab\left(\frac{1+c}{1+a}\right)+bc\left(\frac{1+a}{1+b}\right)+ca\left(\frac{1+b}{1+c}\right)\\ &\geq 3\sqrt[3]{(abc)^2}=3.\end{aligned}$$

Solution 1.44. $\left(\frac{1}{a+b}+\frac{1}{b+c}+\frac{1}{c+a}\right)(a+b+c) \geq \frac{9}{2}$ is equivalent to

$$\left(\frac{1}{a+b}+\frac{1}{b+c}+\frac{1}{c+a}\right)(a+b+b+c+c+a) \geq 9,$$

which follows from Exercise 1.36. For the other inequality use $\frac{1}{a}+\frac{1}{b} \geq \frac{4}{a+b}$. See Exercise 1.22.

Solution 1.45. Note that

$$\frac{n+H_n}{n}=\frac{(1+1)+(1+\frac{1}{2})+\cdots+(1+\frac{1}{n})}{n}.$$

Now, apply the AM-GM inequality.

Solution 1.46. Setting $y_i=\frac{1}{1+x_i}$, then $x_i=\frac{1}{y_i}-1=\frac{1-y_i}{y_i}$. Observe that $y_1+\cdots+y_n=1$ implies that $1-y_i=\sum_{j\neq i} y_i$, then $\sum_{j\neq i} y_i \geq (n-1)\left(\prod_{j\neq i} y_j\right)^{\frac{1}{n-1}}$ and

$$\prod_i x_i=\prod_i\left(\frac{1-y_i}{y_i}\right)=\frac{\prod_i\left(\sum_{j\neq i} y_j\right)}{\prod_i y_i} \geq \frac{(n-1)^n\prod_i\left(\prod_{j\neq i} y_j\right)^{\frac{1}{n-1}}}{\prod_i y_i}=(n-1)^n.$$

Solution 1.47. Define $a_{n+1} = 1-(a_1+\cdots+a_n)$ and $x_i = \frac{1-a_i}{a_i}$ for $i = 1, \ldots, n+1$. Apply Exercise 1.46 directly.

Solution 1.48. $\sum_{i=1}^{n} \frac{1}{1+a_i} = 1 \Rightarrow \sum_{i=1}^{n} \frac{a_i}{1+a_i} = n-1$. Observe that

$$\sum_{i=1}^{n} \sqrt{a_i} - (n-1)\sum_{i=1}^{n} \frac{1}{\sqrt{a_i}} = \sum_{i=1}^{n} \frac{1}{1+a_i} \sum_{i=1}^{n} \sqrt{a_i} - \sum_{i=1}^{n} \frac{a_i}{1+a_i} \sum_{i=1}^{n} \frac{1}{\sqrt{a_i}}$$
$$= \sum_{i,j} \frac{a_i - a_j}{(1+a_j)\sqrt{a_i}} = \sum_{i>j} \frac{(\sqrt{a_i}\sqrt{a_j}-1)(\sqrt{a_i}-\sqrt{a_j})^2(\sqrt{a_i}+\sqrt{a_j})}{(1+a_i)(1+a_j)\sqrt{a_i}\sqrt{a_j}}.$$

Since $1 \geq \frac{1}{1+a_i} + \frac{1}{1+a_j} = \frac{2+a_i+a_j}{1+a_i+a_j+a_ia_j}$, we can deduce that $a_ia_j \geq 1$. Hence the terms of the last sum are positive.

Solution 1.49. Let $S_a = \sum_{i=1}^{n} \frac{a_i^2}{a_i+b_i}$ and $S_b = \sum_{i=1}^{n} \frac{b_i^2}{a_i+b_i}$. Then

$$S_a - S_b = \sum_{i=1}^{n} \frac{a_i^2 - b_i^2}{a_i + b_i} = \sum_{i=1}^{n} a_i - \sum_{i=1}^{n} b_i = 0,$$

thus $S_a = S_b = S$. Hence, we have

$$2S = \sum_{i=1}^{n} \frac{a_i^2 + b_i^2}{a_i + b_i} \geq \frac{1}{2} \sum_{i=1}^{n} \frac{(a_i + b_i)^2}{a_i + b_i} = \sum_{i=1}^{n} a_i,$$

where the inequality follows after using Exercise 1.21.

Solution 1.50. Since the inequality is homogeneous[16] we can assume that $abc = 1$. Setting $x = a^3$, $y = b^3$ and $z = c^3$, the inequality is equivalent to

$$\frac{1}{x+y+1} + \frac{1}{y+z+1} + \frac{1}{z+x+1} \leq 1.$$

Let $A = x+y+1$, $B = y+z+1$ and $C = z+x+1$, then

$$\frac{1}{A} + \frac{1}{B} + \frac{1}{C} \leq 1 \Leftrightarrow (A-1)(B-1)(C-1) - (A+B+C) + 1 \geq 0$$
$$\Leftrightarrow (x+y)(y+z)(z+x) - 2(x+y+z) \geq 2$$
$$\Leftrightarrow (x+y+z)(xy+yz+zx-2) \geq 3.$$

Now, use that

$$\frac{x+y+z}{3} \geq (xzy)^{\frac{1}{3}} \quad \text{and} \quad \frac{xy+yz+zx}{3} \geq (xyz)^{\frac{2}{3}}.$$

[16] A function $f(a, b, \ldots)$ is homogeneous if $f(ta, tb, \ldots) = tf(a, b, \ldots)$ for each $t \in \mathbb{R}$. Then, an inequality of the form $f(a, b, \ldots) \geq 0$, in the case of a homogeneous function, is equivalent to $f(ta, tb, \ldots) \geq 0$ for any $t > 0$.

Second solution. Follow the ideas used in the solution of Exercise 1.15. Start with the inequality $(a^2-b^2)(a-b)\geq 0$ to guarantee that $a^3+b^3+abc\geq ab(a+b+c)$, then

$$\frac{1}{a^3+b^3+abc}\leq\frac{c}{abc(a+b+c)}.$$

Solution 1.51. Note that $abc\leq\left(\frac{a+b+c}{3}\right)^3=\frac{1}{27}$.

$$\begin{aligned}\left(\frac{1}{a}+1\right)\left(\frac{1}{b}+1\right)\left(\frac{1}{c}+1\right)&=1+\frac{1}{a}+\frac{1}{b}+\frac{1}{c}+\frac{1}{ab}+\frac{1}{bc}+\frac{1}{ca}+\frac{1}{abc}\\&\geq 1+\frac{3}{\sqrt[3]{abc}}+\frac{3}{\sqrt[3]{(abc)^2}}+\frac{1}{abc}\\&=\left(1+\frac{1}{\sqrt[3]{abc}}\right)^3\geq 4^3.\end{aligned}$$

Solution 1.52. The inequality is equivalent to $\left(\frac{b+c}{a}\right)\left(\frac{a+c}{b}\right)\left(\frac{a+b}{c}\right)\geq 8$. Now, we use the AM-GM inequality for each term of the product and the inequality follows immediately.

Solution 1.53. Notice that

$$\begin{aligned}&\frac{a}{(a+1)(b+1)}+\frac{b}{(b+1)(c+1)}+\frac{c}{(c+1)(a+1)}\\&=\frac{(a+1)(b+1)(c+1)-2}{(a+1)(b+1)(c+1)}=1-\frac{2}{(a+1)(b+1)(c+1)}\geq\frac{3}{4}\end{aligned}$$

if and only if $(a+1)(b+1)(c+1)\geq 8$, and this last inequality follows immediately from the inequality $\left(\frac{a+1}{2}\right)\left(\frac{b+1}{2}\right)\left(\frac{c+1}{2}\right)\geq\sqrt{a}\sqrt{b}\sqrt{c}=1$.

Solution 1.54. Observe that this exercise is similar to Exercise 1.52.

Solution 1.55. Apply the inequality between the arithmetic mean and the harmonic mean to get

$$\frac{2ab}{a+b}=\frac{2}{\frac{1}{a}+\frac{1}{b}}\leq\frac{a+b}{2}.$$

We can conclude that equality holds when $a=b=c$.

Solution 1.56. First use the fact that $(a+b)^2\geq 4ab$, and then take into account that

$$\sum_{i=1}^{n}\frac{1}{a_ib_i}\geq 4\sum_{i=1}^{n}\frac{1}{(a_i+b_i)^2}.$$

Now, use Exercise 1.36 to prove that

$$\sum_{i=1}^{n}(a_i+b_i)^2\sum_{i=1}^{n}\frac{1}{(a_i+b_i)^2}\geq n^2.$$

Solution 1.57. Using the AM-GM inequality leads to $xy + yz \geq 2y\sqrt{xz}$. Adding similar results we get $2(xy + yz + zx) \geq 2(x\sqrt{yz} + y\sqrt{zx} + z\sqrt{xy})$. Once again, using AM-GM inequality, we get $x^2 + x^2 + y^2 + z^2 \geq 4x\sqrt{yz}$. Adding similar results once more, we obtain $x^2 + y^2 + z^2 \geq x\sqrt{yz} + y\sqrt{zx} + z\sqrt{xy}$. Now adding both results, we reach the conclusion $\frac{(x+y+z)^2}{3} \geq x\sqrt{yz} + y\sqrt{zx} + z\sqrt{xy}$.

Solution 1.58. Using the AM-GM inequality takes us to $x^4 + y^4 \geq 2x^2y^2$. Applying AM-GM inequality once again shows that $2x^2y^2 + z^2 \geq \sqrt{8}xyz$. Or, directly we have that

$$x^4 + y^4 + \frac{z^2}{2} + \frac{z^2}{2} \geq 4\sqrt[4]{\frac{x^4y^4z^4}{4}} = \sqrt{8}xyz.$$

Solution 1.59. Use the AM-GM inequality to obtain

$$\frac{x^2}{y-1} + \frac{y^2}{x-1} \geq 2\frac{xy}{\sqrt{(x-1)(y-1)}} \geq 8.$$

The last inequality follows from $\frac{x}{\sqrt{x-1}} \geq 2$, since $(x-2)^2 \geq 0$.

Second solution. Let $a = x - 1$ and $b = y - 1$, which are positive numbers, then the inequality we need to prove is equivalent to $\frac{(a+1)^2}{b} + \frac{(b+1)^2}{a} \geq 8$. Now, by the AM-GM inequality we have $(a+1)^2 \geq 4a$ and $(b+1)^2 \geq 4b$. Then, $\frac{(a+1)^2}{b} + \frac{(b+1)^2}{a} \geq 4\left(\frac{a}{b} + \frac{b}{a}\right) \geq 8$. The last inequality follows from Exercise 1.24.

Solution 1.60. Observe that (a, b, c) and (a^2, b^2, c^2) have the same order, then use inequality (1.2).

Solution 1.61. By the previous exercise

$$a^3 + b^3 + c^3 \geq a^2b + b^2c + c^2a.$$

Observe that $\left(\frac{1}{a}, \frac{1}{b}, \frac{1}{c}\right)$ and $\left(\frac{1}{a^2}, \frac{1}{b^2}, \frac{1}{c^2}\right)$ can be ordered in the same way. Then, use inequality (1.2) to get

$$\begin{aligned}(ab)^3 + (bc)^3 + (ca)^3 &= \frac{1}{a^3} + \frac{1}{b^3} + \frac{1}{c^3} \\ &\geq \frac{1}{a^2}\frac{1}{c} + \frac{1}{b^2}\frac{1}{a} + \frac{1}{c^2}\frac{1}{b} \\ &= \frac{b}{a} + \frac{c}{b} + \frac{a}{c} \\ &= a^2b + b^2c + c^2a.\end{aligned}$$

Adding these two inequalities leads to the result.

Solution 1.62. Use inequality (1.2) with $(a_1, a_2, a_3) = (b_1, b_2, b_3) = \left(\frac{a}{b}, \frac{b}{c}, \frac{c}{a}\right)$ and $(a_1', a_2', a_3') = \left(\frac{b}{c}, \frac{c}{a}, \frac{a}{b}\right)$.

Solution 1.63. Use inequality (1.2) with $(a_1, a_2, a_3) = (b_1, b_2, b_3) = \left(\frac{1}{a}, \frac{1}{b}, \frac{1}{c}\right)$ and $(a_1', a_2', a_3') = \left(\frac{1}{b}, \frac{1}{c}, \frac{1}{a}\right)$.

Solution 1.64. Assume that $a \leq b \leq c$, and consider $(a_1, a_2, a_3) = (a, b, c)$, then use the rearrangement inequality (1.2) twice over with $(a_1', a_2', a_3') = (b, c, a)$ and (c, a, b), respectively. Note that we are also using

$$(b_1, b_2, b_3) = \left(\frac{1}{b+c-a}, \frac{1}{c+a-b}, \frac{1}{a+b-c}\right).$$

Solution 1.65. Use the same idea as in the previous exercise, but with n variables.

Solution 1.66. Turn to the previous exercise and the fact that $\frac{s}{s-a_1} = 1 + \frac{a_1}{s-a_1}$.

Solution 1.67. Apply Exercise 1.65 to the sequence $a_1, \ldots, a_n, a_1, \ldots, a_n$.

Solution 1.68. Apply Example 1.4.11.

Solution 1.69. Note that $1 = (a^2 + b^2 + c^2) + 2(ab + bc + ca)$, and use the previous exercise as follows:

$$\frac{1}{3} = \frac{a+b+c}{3} \leq \sqrt{\frac{a^2+b^2+c^2}{3}}.$$

Therefore $\frac{1}{3} \leq a^2 + b^2 + c^2$. Hence, $2(ab + bc + ca) \leq \frac{2}{3}$, and the result is evident.

Second solution. The inequality is equivalent to $3(ab + bc + ca) \leq (a + b + c)^2$, which can be simplified to $ab + bc + ca \leq a^2 + b^2 + c^2$.

Solution 1.70. Let $G = \sqrt[n]{x_1 x_2 \cdots x_n}$ be the geometric mean of the given numbers and $(a_1, a_2, \ldots, a_n) = \left(\frac{x_1}{G}, \frac{x_1 x_2}{G^2}, \ldots, \frac{x_1 x_2 \cdots x_n}{G^n}\right)$.

Using Corollary 1.4.2, we can establish that

$$n \leq \frac{a_1}{a_2} + \frac{a_2}{a_3} + \cdots + \frac{a_{n-1}}{a_n} + \frac{a_n}{a_1} = \frac{G}{x_2} + \frac{G}{x_3} + \cdots + \frac{G}{x_n} + \frac{G}{x_1},$$

thus

$$\frac{n}{\frac{1}{x_1} + \cdots + \frac{1}{x_n}} \leq G.$$

Also, using Corollary 1.4.2,

$$n \leq \frac{a_1}{a_n} + \frac{a_2}{a_1} + \cdots + \frac{a_n}{a_{n-1}} = \frac{x_1}{G} + \frac{x_2}{G} + \cdots + \frac{x_n}{G},$$

then

$$G \leq \frac{x_1 + x_2 + \cdots + x_n}{n}.$$

The equalities hold if and only if $a_1 = a_2 = \cdots = a_n$, that is, if and only if $x_1 = x_2 = \cdots = x_n$.

Solution 1.71. The inequality is equivalent to

$$a_1^{n-1} + a_2^{n-1} + \cdots + a_n^{n-1} \geq \frac{a_1 \cdots a_n}{a_1} + \frac{a_1 \cdots a_n}{a_2} + \cdots + \frac{a_1 \cdots a_n}{a_n},$$

which can be verified using the rearrangement inequality several times over.

Solution 1.72. First note that $\sum_{i=1}^n \frac{a_i}{\sqrt{1-a_i}} = \sum_{i=1}^n \frac{1}{\sqrt{1-a_i}} - \sum_{i=1}^n \sqrt{1-a_i}$. Use the AM-GM inequality to obtain

$$\frac{1}{n}\sum_{i=1}^n \frac{1}{\sqrt{1-a_i}} \geq \sqrt[n]{\prod_{i=1}^n \frac{1}{\sqrt{1-a_i}}} = \sqrt{\frac{1}{\sqrt[n]{\prod_{i=1}^n (1-a_i)}}}$$

$$\geq \sqrt{\frac{1}{\frac{1}{n}\sum_{i=1}^n (1-a_i)}} = \sqrt{\frac{n}{n-1}}.$$

Moreover,the Cauchy-Schwarz inequality serves to show that

$$\sum_{i=1}^n \sqrt{1-a_i} \leq \sqrt{\sum_{i=1}^n (1-a_i)}\sqrt{n} = \sqrt{n(n-1)} \quad \text{and} \quad \sum_{i=1}^n \sqrt{a_i} \leq \sqrt{n}.$$

Solution 1.73. (i) $\sqrt{4a+1} < \frac{4a+1+1}{2} = 2a+1$.
(ii) Use the Cauchy-Schwarz inequality with $u = (\sqrt{4a+1}, \sqrt{4b+1}, \sqrt{4c+1})$ and $v = (1,1,1)$.

Solution 1.74. Suppose that $a \geq b \geq c \geq d$ (the other cases are similar). Then, if $A = b+c+d$, $B = a+c+d$, $C = a+b+d$ and $D = a+b+c$, we can deduce that $\frac{1}{A} \geq \frac{1}{B} \geq \frac{1}{C} \geq \frac{1}{D}$. Apply the Tchebyshev inequality twice over to show that

$$\frac{a^3}{A} + \frac{b^3}{B} + \frac{c^3}{C} + \frac{d^3}{D} \geq \frac{1}{4}(a^3+b^3+c^3+d^3)\left(\frac{1}{A}+\frac{1}{B}+\frac{1}{C}+\frac{1}{D}\right)$$

$$\geq \frac{1}{16}(a^2+b^2+c^2+d^2)(a+b+c+d)\left(\frac{1}{A}+\frac{1}{B}+\frac{1}{C}+\frac{1}{D}\right)$$

$$= \frac{1}{16}(a^2+b^2+c^2+d^2)\left(\frac{A+B+C+D}{3}\right)\left(\frac{1}{A}+\frac{1}{B}+\frac{1}{C}+\frac{1}{D}\right).$$

Now, use the Cauchy-Schwarz inequality to derive the result

$$a^2+b^2+c^2+d^2 \geq ab+bc+cd+da = 1$$

and the inequality $(A+B+C+D)(\frac{1}{A}+\frac{1}{B}+\frac{1}{C}+\frac{1}{D}) \geq 16$.

Solution 1.75. Apply the rearrangement inequality to

$$(a_1, a_2, a_3) = \left(\sqrt[3]{\frac{a}{b}}, \sqrt[3]{\frac{b}{c}}, \sqrt[3]{\frac{c}{a}}\right), \ (b_1, b_2, b_3) = \left(\sqrt[3]{\left(\frac{a}{b}\right)^2}, \sqrt[3]{\left(\frac{b}{c}\right)^2}, \sqrt[3]{\left(\frac{c}{a}\right)^2}\right)$$

and the permutation $(a_1', a_2', a_3') = \left(\sqrt[3]{\frac{b}{c}}, \sqrt[3]{\frac{c}{a}}, \sqrt[3]{\frac{a}{b}}\right)$ to derive

$$\frac{a}{b} + \frac{b}{c} + \frac{c}{a} \geq \sqrt[3]{\frac{a^2}{bc}} + \sqrt[3]{\frac{b^2}{ca}} + \sqrt[3]{\frac{c^2}{ab}}.$$

Finally, use the fact that $abc = 1$.

Second solution. The AM-GM inequality and the fact that $abc = 1$ imply that

$$\frac{1}{3}\left(\frac{a}{b} + \frac{a}{b} + \frac{b}{c}\right) \geq \sqrt[3]{\frac{a}{b}\frac{a}{b}\frac{b}{c}} = \sqrt[3]{\frac{a^2}{bc}} = \sqrt[3]{\frac{a^3}{abc}} = a.$$

Similarly,

$$\frac{1}{3}\left(\frac{b}{c} + \frac{b}{c} + \frac{c}{a}\right) \geq b \text{ and } \frac{1}{3}\left(\frac{c}{a} + \frac{c}{a} + \frac{a}{b}\right) \geq c,$$

and the result follows.

Solution 1.76. Using the hypothesis, for all k, leads to $s - 2x_k > 0$. Turn to the Cauchy-Schwarz inequality to show that

$$\left(\sum_{k=1}^{n} \frac{x_k^2}{s - 2x_k}\right)\left(\sum_{k=1}^{n}(s - 2x_k)\right) \geq \left(\sum_{k=1}^{n} x_k\right)^2 = s^2.$$

But $0 < \sum_{k=1}^{n}(s - 2x_k) = ns - 2s$, therefore

$$\sum_{k=1}^{n} \frac{x_k^2}{s - 2x_k} \geq \frac{s}{n-2}.$$

Solution 1.77. The function $f(x) = \left(x + \frac{1}{x}\right)^2$ is convex in $\mathbb{R}^+$.

Solution 1.78. The function

$$f(a,b,c) = \frac{a}{b+c+1} + \frac{b}{a+c+1} + \frac{c}{a+b+1} + (1-a)(1-b)(1-c)$$

is convex in each variable, therefore its maximum is attained at the endpoints.

Solution 1.79. If $x = 0$, then the inequality reduces to $1 + \frac{1}{\sqrt{1+y^2}} \leq 2$, which is true because $y \geq 0$. By symmetry, the inequality holds for $y = 0$.

Now, suppose that $0 < x \leq 1$ and $0 < y \leq 1$. Let $u \geq 0$ and $v \geq 0$ such that $x = e^{-u}$ and $y = e^{-v}$, then the inequality becomes

$$\frac{1}{\sqrt{1+e^{-2u}}} + \frac{1}{\sqrt{1+e^{-2v}}} \leq \frac{2}{\sqrt{1+e^{-(u+v)}}},$$

that is,

$$\frac{f(u)+f(v)}{2} \leq f\left(\frac{u+v}{2}\right),$$

where $f(x) = \frac{1}{\sqrt{1+e^{-2x}}}$. Since $f''(x) = \frac{1-2e^{2x}}{(1+e^{-2x})^{5/2}e^{4x}}$, the function is concave in the interval $[0,\infty)$. Thus the previous inequality holds.

Solution 1.80. Find $f''(x)$.

Solution 1.81. Use $\log(\sin x)$ or the fact that

$$\sin A \sin B = \sin\left(\frac{A+B}{2}+\frac{A-B}{2}\right)\sin\left(\frac{A+B}{2}-\frac{A-B}{2}\right).$$

Solution 1.82. (i) If $1+nx \leq 0$, the inequality is evident since $(1+x)^n \geq 0$. Suppose that $(1+nx) > 0$. Apply AM-GM inequality to the numbers $(1,1,\ldots,1,1+nx)$ with $(n-1)$ ones.
(ii) Let $a_1,\ldots, a_n$ be positive numbers and define, for each $j = 1,\ldots n$, $\sigma_j = \frac{a_1+\cdots+a_j}{j}$. Apply Bernoulli's inequality to show that $\left(\frac{\sigma_j}{\sigma_{j-1}}\right)^j \geq j\frac{\sigma_j}{\sigma_{j-1}} - (j-1)$, which implies

$$\sigma_j^j \geq \sigma_{j-1}^j\left(j\frac{\sigma_j}{\sigma_{j-1}} - (j-1)\right) = \sigma_{j-1}^{j-1}(j\sigma_j - (j-1)\sigma_{j-1}) = a_j\sigma_{j-1}^{j-1}.$$

Then, $\sigma_n^n \geq a_n\sigma_{n-1}^{n-1} \geq a_na_{n-1}\sigma_{n-2}^{n-2} \geq \cdots \geq a_na_{n-1}\cdots a_1$.

Solution 1.83. If $x \geq y \geq z$, we have $x^n(x-y)(x-z) \geq y^n(x-y)(y-z)$ and $z^n(z-x)(z-y) \geq 0$.

Solution 1.84. Notice that $x(x-z)^2+y(y-z)^2-(x-z)(y-z)(x+y-z) \geq 0$ if and only if $x(x-z)(x-y)+y(y-z)(y-x)+z(x-z)(y-z) \geq 0$. The inequality now follows from Schür's inequality. Alternatively, we can see that the last expression is symmetric in x, y and z, then we can assume $x \geq z \geq y$, and if we return to the original inequality, it becomes clear that

$$x(x-z)^2+y(y-z)^2 \geq 0 \geq (x-z)(y-z)(x+y-z).$$

Solution 1.85. The inequality is homogeneous, therefore we can assume that $a+b+c=1$. Now, the terms on the left-hand side are of the form $\frac{x}{(1-x)^2}$ and the function $f(x) = \frac{x}{(1-x)^2}$ is convex, since $f''(x) = \frac{4+2x}{(1-x)^4} > 0$. By Jensen's inequality it follows that $\frac{a}{(1-a)^2}+\frac{b}{(1-b)^2}+\frac{c}{(1-c)^2} \geq 3f\left(\frac{a+b+c}{3}\right) = 3f\left(\frac{1}{3}\right) = \left(\frac{3}{2}\right)^2$.

Solution 1.86. Since $(a+b+c)^2 \geq 3(ab+bc+ca)$, we can deduce that $1+\frac{3}{ab+bc+ca} \geq 1+\frac{9}{(a+b+c)^2}$. Thus, the inequality will hold if

$$1+\frac{9}{(a+b+c)^2} \geq \frac{6}{(a+b+c)}.$$

But this last inequality follows from $\left(1-\frac{3}{a+b+c}\right)^2 \geq 0$.

Now, if $abc = 1$, consider $x = \frac{1}{a}$, $y = \frac{1}{b}$ and $z = \frac{1}{c}$; it follows immediately that $xyz = 1$. Thus, the inequality is equivalent to

$$1 + \frac{3}{xy + yz + zx} \geq \frac{6}{x + y + z}$$

which is the first part of this exercise.

Solution 1.87. We will use the convexity of the function $f(x) = x^r$ for $r \geq 1$ (its second derivative is $r(r-1)x^{r-2}$). First suppose that $r > s > 0$. Then Jensen's inequality for the convex function $f(x) = x^{\frac{r}{s}}$ applied to $x_1^s, \ldots, x_n^s$ gives

$$t_1 x_1^r + \cdots + t_n x_n^r \geq (t_1 x_1^s + \cdots + t_n x_n^s)^{\frac{r}{s}}$$

and taking the $\frac{1}{r}$-th power of both sides gives the desired inequality.

Now suppose $0 > r > s$. Then $f(x) = x^{\frac{r}{s}}$ is concave, so Jensen's inequality is reversed; however, taking $\frac{1}{r}$-th powers reverses the inequality again.

Finally, in the case $r > 0 > s$, $f(x) = x^{\frac{r}{s}}$ is again convex, and taking $\frac{1}{r}$-th powers preserves the inequality.

Solution 1.88. (i) Apply Hölder's inequality to the numbers $x_1^c, \ldots, x_n^c, y_1^c, \ldots, y_n^c$ with $a' = \frac{a}{c}$ and $b' = \frac{b}{c}$.
(ii) Proceed as in Example 1.5.9. The only extra fact that we need to prove is $x_i y_i z_i \leq \frac{1}{a}x_i^a + \frac{1}{b}y_i^b + \frac{1}{c}z_i^c$, but this follows from part (i) of that example.

Solution 1.89. By the symmetry of the variables in the inequality we can assume that $a \leq b \leq c$. We have two cases, (i) $b \leq \frac{a+b+c}{3}$ and (ii) $b \geq \frac{a+b+c}{3}$.
Case (i): $b \leq \frac{a+b+c}{3}$.
It happens that $\frac{a+b+c}{3} \leq \frac{a+c}{2} \leq c$, and it is true that $\frac{a+b+c}{3} \leq \frac{b+c}{2} \leq c$. Then, there exist λ, $\mu \in [0, 1]$ such that

$$\frac{c+a}{2} = \lambda c + (1-\lambda)\left(\frac{a+b+c}{3}\right) \quad \text{and} \quad \frac{b+c}{2} = \mu c + (1-\mu)\left(\frac{a+b+c}{3}\right).$$

Adding these equalities, we obtain

$$\frac{a+b+2c}{2} = (\lambda+\mu)c + (2-\lambda-\mu)\left(\frac{a+b+c}{3}\right) = (2-\lambda-\mu)\left(\frac{a+b-2c}{3}\right) + 2c.$$

Hence,

$$\frac{a+b-2c}{2} = (2-\lambda-\mu)\left(\frac{a+b-2c}{3}\right),$$

therefore $2-(\lambda+\mu) = \frac{3}{2}$ and $(\lambda+\mu) = \frac{1}{2}$.

Now, since f is a convex function, we have

$$f\left(\frac{a+b}{2}\right) \leq \frac{1}{2}\left(f(a)+f(b)\right)$$
$$f\left(\frac{b+c}{2}\right) \leq \mu f(c)+(1-\mu) f\left(\frac{a+b+c}{3}\right)$$
$$f\left(\frac{c+a}{2}\right) \leq \lambda f(c)+(1-\lambda) f\left(\frac{a+b+c}{3}\right)$$

thus, adding these inequalities we get

$$f\left(\frac{a+b}{2}\right)+f\left(\frac{b+c}{2}\right)+f\left(\frac{c+a}{2}\right) \leq \frac{1}{2}\left(f(a)+f(b)+f(c)\right) + \frac{3}{2} f\left(\frac{a+b+c}{3}\right).$$

Case (ii): $b \geq \frac{a+b+c}{3}$.
It is similar to case (i), using the fact that $a \leq \frac{a+c}{2} \leq \frac{a+b+c}{3}$ and $a \leq \frac{a+b}{2} \leq \frac{a+b+c}{3}$.

Solution 1.90. If any of a, b or c is zero, the inequality is evident. Applying Popoviciu's inequality (see the previous exercise) to the function $f : \mathbb{R} \to \mathbb{R}^+$ defined by $f(x) = \exp(2x)$, which is convex since $f''(x) = 4\exp(2x) > 0$, we obtain

$$\begin{aligned} &\exp(2x)+\exp(2y)+\exp(2z)+3\exp\left(\frac{2(x+y+z)}{3}\right) \\ &\geq 2\left[\exp(x+y)+\exp(y+z)+\exp(z+x)\right] \\ &= 2\left[\exp(x)\exp(y)+\exp(y)\exp(z)+\exp(z)\exp(x)\right]. \end{aligned}$$

Setting $a = \exp(x)$, $b = \exp(y)$, $c = \exp(z)$, the previous inequality can be rewritten as

$$a^2+b^2+c^2+3\sqrt[3]{a^2b^2c^2} \geq 2(ab+bc+ca).$$

For the second part apply the AM-GM inequality in the following way:

$$2abc+1 = abc+abc+1 \geq 3\sqrt[3]{a^2b^2c^2}.$$

Solution 1.91. Apply Popoviciu's inequality to the convex function $f(x) = x + \frac{1}{x}$. We will get the inequality $\frac{1}{a}+\frac{1}{b}+\frac{1}{c}+\frac{9}{a+b+c} \geq \frac{4}{b+c}+\frac{4}{c+a}+\frac{4}{a+b}$. Then multiply both sides by $(a+b+c)$ to finish the proof.

Solution 1.92. Observe that by using (1.8), we obtain

$$x^2+y^2+z^2-|x||y|-|y||z|-|z||x| = \frac{1}{2}(|x|-|y|)^2+\frac{1}{2}(|y|-|z|)^2+\frac{1}{2}(|z|-|x|)^2,$$

which is clearly greater than or equal to zero. Hence

$$|xy+yz+zx| \leq |x||y|+|y||z|+|z||x| \leq x^2+y^2+z^2.$$

Second solution. Apply Cauchy-Schwarz inequality to (x, y, z) and (y, z, x).

Solution 1.93. The inequality is equivalent to $ab + bc + ca \leq a^2 + b^2 + c^2$, which we know is true. See Exercise 1.27.

Solution 1.94. Observe that if $a + b + c = 0$, then it follows from (1.7) that $a^3 + b^3 + c^3 = 3abc$. Since $(x - y) + (y - z) + (z - x) = 0$, we can derive the following factorization:

$$(x - y)^3 + (y - z)^3 + (z - x)^3 = 3(x - y)(y - z)(z - x).$$

Solution 1.95. Assume, without loss of generality, that $a \geq b \geq c$. We need to prove that

$$-a^3 + b^3 + c^3 + 3abc \geq 0.$$

Since

$$-a^3 + b^3 + c^3 + 3abc = (-a)^3 + b^3 + c^3 - 3(-a)bc,$$

the latter expression factors into

$$\frac{1}{2}(-a + b + c)((a + b)^2 + (a + c)^2 + (b - c)^2).$$

The conclusion now follows from the triangle inequality, $b + c > a$.

Solution 1.96. Let $p = |(x - y)(y - z)(z - x)|$. Using AM-GM inequality on the right-hand side of identity (1.8), we get

$$x^2 + y^2 + z^2 - xy - yz - zx \geq \frac{3}{2}\sqrt[3]{p^2}. \tag{4.1}$$

Now, since $|x - y| \leq x + y$, $|y - z| \leq y + z$, $|z - x| \leq z + x$, it follows that

$$2(x + y + z) \geq |x - y| + |y - z| + |z - x|. \tag{4.2}$$

Applying again the AM-GM inequality leads to

$$2(x + y + z) \geq 3\sqrt[3]{p},$$

and the result follows from inequalities (4.1) and (4.2).

Solution 1.97. Using identity (1.7), the condition $x^3 + y^3 + z^3 - 3xyz = 1$ can be factorized as

$$(x + y + z)(x^2 + y^2 + z^2 - xy - yz - zx) = 1. \tag{4.3}$$

Let $A = x^2 + y^2 + z^2$ and $B = x + y + z$. Notice that $B^2 - A = 2(xy + yz + zx)$. By identity (1.8), we have that $B > 0$. Equation (4.3) now becomes

$$B\left(A - \frac{B^2 - A}{2}\right) = 1,$$

therefore $3A = B^2 + \frac{2}{B}$. Since $B > 0$, we may apply the AM-GM inequality to obtain

$$3A = B^2 + \frac{2}{B} = B^2 + \frac{1}{B} + \frac{1}{B} \geq 3,$$

that is, $A \geq 1$. For instance, the minimum $A = 1$ is attained when $(x, y, z) = (1, 0, 0)$.

Solution 1.98. Inequality (1.11) helps to establish

$$\frac{1}{a}+\frac{1}{b}+\frac{4}{c}+\frac{16}{d} \geq \frac{(1+1+2+4)^2}{a+b+c+d} = \frac{64}{a+b+c+d}.$$

Solution 1.99. Apply inequality (1.11) twice over to get

$$a^4+b^4 = \frac{a^4}{1}+\frac{b^4}{1} \geq \frac{(a^2+b^2)^2}{2} \geq \frac{(\frac{(a+b)^2}{2})^2}{2} = \frac{(a+b)^4}{8}.$$

Solution 1.100. Express the left-hand side as

$$\frac{(\sqrt{2})^2}{x+y}+\frac{(\sqrt{2})^2}{y+z}+\frac{(\sqrt{2})^2}{z+x}$$

and use inequality (1.11).

Solution 1.101. Express the left-hand side as

$$\frac{x^2}{axy+bzx}+\frac{y^2}{ayz+bxy}+\frac{z^2}{azx+byz},$$

and then use inequality (1.11) to get

$$\frac{x^2}{axy+bzx}+\frac{y^2}{ayz+bxy}+\frac{z^2}{azx+byz} \geq \frac{(x+y+z)^2}{(a+b)(xy+yz+zx)} \geq \frac{3}{a+b},$$

where the last inequality follows from (1.8).

Solution 1.102. Rewrite the left-hand side as

$$\frac{a^2}{a+b}+\frac{b^2}{b+c}+\frac{c^2}{a+c}+\frac{b^2}{a+b}+\frac{c^2}{b+c}+\frac{a^2}{a+c},$$

and then apply inequality (1.11).

Solution 1.103. (i) Express the left-hand side as

$$\frac{x^2}{x^2+2xy+3zx}+\frac{y^2}{y^2+2yz+3xy}+\frac{z^2}{z^2+2zx+3yz}$$

and apply inequality (1.11) to get

$$\frac{x}{x+2y+3z}+\frac{y}{y+2z+3x}+\frac{z}{z+2x+3y} \geq \frac{(x+y+z)^2}{x^2+y^2+z^2+5(xy+yz+zx)}.$$

Now it suffices to prove that

$$\frac{(x+y+z)^2}{x^2+y^2+z^2+5(xy+yz+zx)} \geq \frac{1}{2},$$

but this is equivalent to $x^2 + y^2 + z^2 \geq xy + yz + zx$.
(ii) Proceed as in part (i), expressing the left-hand side as

$$\frac{w^2}{xw + 2yw + 3zw} + \frac{x^2}{xy + 2xz + 3xw} + \frac{y^2}{yz + 2yw + 3xy} + \frac{z^2}{zw + 2xz + 3yz},$$

then use inequality (1.11) to get

$$\frac{w}{x + 2y + 3z} + \frac{x}{y + 2z + 3w} + \frac{y}{z + 2w + 3x} + \frac{z}{w + 2x + 3y} \\ \geq \frac{(w + x + y + z)^2}{4(wx + xy + yz + zw + wy + xz)}.$$

Then, the inequality we have to prove becomes

$$\frac{(w + x + y + z)^2}{4(wx + xy + yz + zw + wy + xz)} \geq \frac{2}{3},$$

which is equivalent to $3(w^2+x^2+y^2+z^2) \geq 2(wx+xy+yz+zw+wy+xz)$. This follows by using the AM-GM inequality six times under the form $x^2 + y^2 \geq 2xy$.

Solution 1.104. We again apply inequality (1.11) to get

$$\frac{x^2}{(x + y)(x + z)} + \frac{y^2}{(y + z)(y + x)} + \frac{z^2}{(z + x)(z + y)} \\ \geq \frac{(x + y + z)^2}{x^2 + y^2 + z^2 + 3(xy + yz + zx)}.$$

Also, the inequality

$$\frac{(x + y + z)^2}{x^2 + y^2 + z^2 + 3(xy + yz + zx)} \geq \frac{3}{4}$$

is equivalent to

$$x^2 + y^2 + z^2 \geq xy + yz + zx.$$

Solution 1.105. We express the left-hand side as

$$\frac{a^2}{a(b + c)} + \frac{b^2}{b(c + d)} + \frac{c^2}{c(d + a)} + \frac{d^2}{d(a + b)}$$

and apply inequality (1.11) to get

$$\frac{a^2}{a(b + c)} + \frac{b^2}{b(c + d)} + \frac{c^2}{c(d + a)} + \frac{d^2}{d(a + b)} \geq \frac{(a + b + c + d)^2}{a(b + 2c + d) + b(c + d) + d(b + c)}.$$

On the other hand, observe that

$$\frac{(a+b+c+d)^2}{(ac+bd)+(ab+ac+ad+bc+bd+cd)}$$
$$= \frac{a^2+b^2+c^2+d^2+2ab+2ac+2ad+2bc+2bd+2cd}{(ac+bd)+(ab+ac+ad+bc+bd+cd)}.$$

To prove that this last expression is greater than 2 is equivalent to showing that $a^2+c^2 \geq 2ac$ and $b^2+d^2 \geq 2bd$, which can be done using the AM-GM inequality.

Solution 1.106. We express the left-hand side as

$$\frac{a^2}{ab+ac} + \frac{b^2}{bc+bd} + \frac{c^2}{cd+ce} + \frac{d^2}{de+ad} + \frac{e^2}{ae+be}$$

and apply inequality (1.11) to get

$$\frac{a^2}{ab+ac} + \frac{b^2}{bc+bd} + \frac{c^2}{cd+ce} + \frac{d^2}{de+ad} + \frac{e^2}{ae+be} \geq \frac{(a+b+c+d+e)^2}{\sum ab}.$$

Since

$$(a+b+c+d+e)^2 = \sum a^2 + 2\sum ab,$$

we have to prove that

$$2\sum a^2 + 4\sum ab \geq 5\sum ab,$$

which is equivalent to

$$2\sum a^2 \geq \sum ab.$$

The last inequality follows from $\sum a^2 \geq \sum ab$.

Solution 1.107. (i) Using Tchebyshev's inequality with the collections $(a \geq b \geq c)$ and $(\frac{a^2}{x} \geq \frac{b^2}{y} \geq \frac{c^2}{z})$, we obtain

$$\frac{1}{3}\left(\frac{a^3}{x} + \frac{b^3}{y} + \frac{c^3}{z}\right) \geq \frac{\frac{a^2}{x} + \frac{b^2}{y} + \frac{c^2}{z}}{3} \cdot \frac{a+b+c}{3},$$

then by (1.11), we can deduce that

$$\frac{a^2}{x} + \frac{b^2}{y} + \frac{c^2}{z} \geq \frac{(a+b+c)^2}{x+y+z}.$$

Therefore

$$\frac{a^3}{x} + \frac{b^3}{y} + \frac{c^3}{z} \geq \frac{(a+b+c)^2}{x+y+z} \cdot \frac{a+b+c}{3}.$$

(ii) By Exercise 1.88, we have

$$\left(\frac{a^3}{x}+\frac{b^3}{y}+\frac{c^3}{z}\right)^{\frac{1}{3}}(1+1+1)^{\frac{1}{3}}(x+y+z)^{\frac{1}{3}} \geq a+b+c.$$

Raising to the cubic power both sides and then dividing both sides by $3(x+y+z)$ we obtain the result.

Solution 1.108. Using inequality (1.11), we obtain

$$\begin{aligned}
&\frac{x_1^2+x_2^2+\cdots+x_n^2}{x_1+x_2+\cdots+x_n}\\
&= \frac{x_1^2}{x_1+x_2+\cdots+x_n}+\frac{x_2^2}{x_1+x_2+\cdots+x_n}+\cdots+\frac{x_n^2}{x_1+x_2+\cdots+x_n}\\
&\geq \frac{(x_1+x_2+\cdots+x_n)^2}{n(x_1+x_2+\cdots+x_n)} = \frac{x_1+x_2+\cdots+x_n}{n}.
\end{aligned}$$

Thus, it is enough to prove that

$$\left(\frac{x_1+x_2+\cdots+x_n}{n}\right)^{\frac{kn}{t}} \geq x_1\cdot x_2\cdot\cdots\cdot x_n.$$

Since $k=\max\{x_1,x_2,\ldots,x_n\} \geq \min\{x_1,x_2,\ldots,x_n\}=t$, we have that $\frac{kn}{t}\geq n$ and since $\frac{x_1+x_2+\cdots+x_n}{n}\geq 1$, because all the x_i are positive integers, it is enough to prove that

$$\left(\frac{x_1+x_2+\cdots+x_n}{n}\right)^{n} \geq x_1\cdot x_2\cdot\cdots\cdot x_n,$$

which is equivalent to the AM-GM inequality.

Because all the intermediate inequalities are valid as equalities when $x_1 = x_2=\cdots=x_n$, we conclude that equality happens when $x_1=x_2=\cdots=x_n$.

Solution 1.109. Using the substitution $a=\frac{x}{y}$, $b=\frac{y}{z}$ and $c=\frac{z}{x}$, the inequality takes the form

$$\frac{a^3}{a^3+2}+\frac{b^3}{b^3+2}+\frac{c^3}{c^3+2}\geq 1,$$

and with the extra condition, $abc=1$.

In order to prove this last inequality the extra condition is used as follows:

$$\begin{aligned}
\frac{a^3}{a^3+2}+\frac{b^3}{b^3+2}+\frac{c^3}{c^3+2} &= \frac{a^3}{a^3+2abc}+\frac{b^3}{b^3+2abc}+\frac{c^3}{c^3+2abc}\\
&= \frac{a^2}{a^2+2bc}+\frac{b^2}{b^2+2ca}+\frac{c^2}{c^2+2ab}\\
&\geq \frac{(a+b+c)^2}{a^2+b^2+c^2+2bc+2ca+2ab}=1.
\end{aligned}$$

The inequality above follows from inequality (1.11).

Solution 1.110. With the substitution $x = \frac{a}{b}$, $y = \frac{b}{c}$, $z = \frac{c}{a}$, the inequality takes the form

$$\frac{a}{b+c} + \frac{b}{c+a} + \frac{c}{a+b} \geq \frac{3}{2},$$

which is Nesbitt's inequality (Example 1.4.8).

Solution 1.111. Use the substitution $x_1 = \frac{a_2}{a_1}$, $x_2 = \frac{a_3}{a_2}$, ..., $x_n = \frac{a_1}{a_n}$. Since $\frac{1}{1+x_1+x_1x_2} = \frac{1}{1+\frac{a_2}{a_1}+\frac{a_2}{a_1}\frac{a_3}{a_2}} = \frac{a_1}{a_1+a_2+a_3}$ and similarly for the other terms on the left-hand side of the inequality, the inequality we have to prove becomes

$$\frac{a_1}{a_1 + a_2 + a_3} + \frac{a_2}{a_2 + a_3 + a_4} + \cdots + \frac{a_n}{a_n + a_1 + a_2} > 1.$$

But this inequality is easy to prove. It is enough to observe that for all $i = 1, \ldots, n$ we have

$$a_i + a_{i+1} + a_{i+2} < a_1 + a_2 + \cdots + a_n.$$

Solution 1.112. Using the substitution $x = \frac{1}{a}$, $y = \frac{1}{b}$, $z = \frac{1}{c}$, the condition $ab + bc + ca = abc$ becomes $x + y + z = 1$ and the inequality is equivalent to

$$\frac{x^4 + y^4}{x^3 + y^3} + \frac{y^4 + z^4}{y^3 + z^3} + \frac{z^4 + x^4}{z^3 + x^3} \geq 1 = x + y + z.$$

Tchebyshev's inequality can be used to prove that

$$\frac{x^4 + y^4}{2} \geq \frac{x^3 + y^3}{2}\frac{x + y}{2},$$

thus

$$\frac{x^4 + y^4}{x^3 + y^3} + \frac{y^4 + z^4}{y^3 + z^3} + \frac{z^4 + x^4}{z^3 + x^3} \geq \frac{x + y}{2} + \frac{y + z}{2} + \frac{z + x}{2}.$$

Solution 1.113. The inequality on the right-hand side follows from inequality (1.11). For the inequality on the left-hand side, the substitution $x = \frac{bc}{a}$, $y = \frac{ca}{b}$, $z = \frac{ab}{c}$ transforms the inequality into

$$\frac{x + y + z}{3} \geq \sqrt{\frac{yz + zx + xy}{3}}.$$

Squaring both sides, we obtain $3(xy + yz + zx) \leq (x + y + z)^2$, which is valid if and only if $(xy + yz + zx) \leq x^2 + y^2 + z^2$, something we already know.

Solution 1.114. Note that

$$\frac{a-2}{a+1} + \frac{b-2}{b+1} + \frac{c-2}{c+1} \leq 0 \Leftrightarrow 3 - 3\left(\frac{1}{a+1} + \frac{1}{b+1} + \frac{1}{c+1}\right) \leq 0$$
$$\Leftrightarrow 1 \leq \frac{1}{a+1} + \frac{1}{b+1} + \frac{1}{c+1}.$$

Using the substitution $a = \frac{2x}{y}$, $b = \frac{2y}{z}$, $c = \frac{2z}{x}$, we get

$$\begin{aligned}\frac{1}{a+1} + \frac{1}{b+1} + \frac{1}{c+1} &= \frac{1}{\frac{2x}{y}+1} + \frac{1}{\frac{2y}{z}+1} + \frac{1}{\frac{2z}{x}+1}\\ &= \frac{y}{2x+y} + \frac{z}{2y+z} + \frac{x}{2z+x}\\ &= \frac{y^2}{2xy+y^2} + \frac{z^2}{2yz+z^2} + \frac{x^2}{2zx+x^2}\\ &\geq \frac{(x+y+z)^2}{2xy+y^2+2yz+z^2+2zx+x^2} = 1.\end{aligned}$$

The only inequality in the expression follows from inequality (1.11).

Solution 1.115. Observe that

$$[5,0,0] = \frac{2}{6}(a^5+b^5+c^5) \geq \frac{2}{6}(a^3bc+b^3ca+c^3ab) = [3,1,1],$$

where Muirhead's theorem has been used.

Solution 1.116. Using Heron's formula for the area of a triangle, we can rewrite the inequality as

$$a^2+b^2+c^2 \geq 4\sqrt{3}\sqrt{\frac{(a+b+c)}{2}\frac{(a+b-c)}{2}\frac{(a+c-b)}{2}\frac{(b+c-a)}{2}}.$$

This is equivalent to

$$\begin{aligned}(a^2+b^2+c^2)^2 &\geq 3[((a+b)^2-c^2)(c^2-(b-a)^2)]\\ &= 3(2c^2a^2+2c^2b^2+2a^2b^2-(a^4+b^4+c^4)),\end{aligned}$$

that is, $a^4+b^4+c^4 \geq a^2b^2+b^2c^2+c^2a^2$, which, in terms of Muirhead's theorem, is equivalent to proving $[4,0,0] \geq [2,2,0]$.

Second solution. Using the substitution

$$x = a+b-c,\ \ y = a-b+c,\ \ z = -a+b+c,$$

we obtain $x+y+z = a+b+c$; then, using Heron's formula we get

$$4(ABC) = \sqrt{(a+b+c)(xyz)} \leq \sqrt{(a+b+c)\frac{(x+y+z)^3}{27}} = \frac{(a+b+c)^2}{3\sqrt{3}}.$$

Now we only need to prove that $(a+b+c)^2 \leq 3(a^2+b^2+c^2)$. This last inequality follows from Muirhead's theorem, since $[1,1,0] \leq [2,0,0]$.

Solution 1.117. Notice that

$$\begin{aligned}&\frac{a}{(a+b)(a+c)}+\frac{b}{(b+c)(b+a)}+\frac{c}{(c+a)(c+b)}\leq\frac{9}{4(a+b+c)}\\&\Leftrightarrow 8(ab+bc+ca)(a+b+c)\leq 9(a+b)(b+c)(c+a)\\&\Leftrightarrow 24abc+8\sum(a^2b+ab^2)\leq 9\sum(a^2b+ab^2)+18abc\\&\Leftrightarrow 6abc\leq a^2b+ab^2+b^2c+bc^2+c^2a+ca^2\\&\Leftrightarrow [1,1,1]\leq[2,1,0]\,.\end{aligned}$$

Solution 1.118. The inequality is equivalent to

$$a^3+b^3+c^3\geq ab(a+b-c)+bc(b+c-a)+ca(c+a-b).$$

Setting $x=a+b-c$, $y=b+c-a$, $z=a+c-b$, we get $a=\frac{z+x}{2}$, $b=\frac{x+y}{2}$, $c=\frac{y+z}{2}$. Then, the inequality we have to prove is

$$\frac{1}{8}((z+x)^3+(x+y)^3+(y+z)^3)\geq\frac{1}{4}((z+x)(x+y)x+(x+y)(y+z)y+(y+z)(z+x)z),$$

which is again equivalent to

$$3(x^2y+y^2x+\cdots+z^2x)\geq 2(x^2y+\cdots)+6xyz$$

or

$$x^2y+y^2x+y^2z+z^2y+z^2x+x^2z\geq 6xyz,$$

and applying Muirhead's theorem we obtain the result when x, y, z are non-negative. If one of them is negative (and it cannot be more than one at a time), we will get

$$x^2(y+z)+y^2(z+x)+z^2(x+y)=x^2 2c+y^2 2a+z^2 2b\geq 0$$

but $6xyz$ is negative, which ends the proof.

Solution 1.119. Observe that

$$\frac{a^3}{b^2-bc+c^2}+\frac{b^3}{c^2-ca+a^2}+\frac{c^3}{a^2-ab+b^2}\geq a+b+c$$

is equivalent to the inequality

$$\frac{a^3(b+c)}{b^3+c^3}+\frac{b^3(c+a)}{c^3+a^3}+\frac{c^3(a+b)}{a^3+b^3}\geq a+b+c,$$

which in turn is equivalent to

$$\begin{aligned}&a^3(b+c)(a^3+c^3)(a^3+b^3)+b^3(c+a)(b^3+c^3)(a^3+b^3)\\&\quad+c^3(a+b)(a^3+c^3)(b^3+c^3)\\&\geq(a+b+c)(a^3+b^3)(b^3+c^3)(c^3+a^3).\end{aligned}$$

The last inequality can be written in the terminology of Muirhead's theorem as

$$\begin{aligned}[9,1,0]+[6,4,0]+[6,3,1]+[4,3,3] &\geq \left(\frac{1}{2}[1,0,0]\right)\left([6,3,0]+\frac{1}{3}[3,3,3]\right)\\ &= [7,3,0]+[6,4,0]+[6,3,1]+[4,3,3]\\ &\Leftrightarrow [9,1,0]\geq[7,3,0],\end{aligned}$$

a direct result of Muirhead's theorem.

Solution 1.120. Suppose that $a \leq b \leq c$, then

$$\frac{1}{(1+b)(1+c)} \leq \frac{1}{(1+c)(1+a)} \leq \frac{1}{(1+a)(1+b)}.$$

Use Tchebyshev's inequality to prove that

$$\begin{aligned}&\frac{a^3}{(1+b)(1+c)}+\frac{b^3}{(1+c)(1+a)}+\frac{c^3}{(1+a)(1+b)}\\ &\geq \frac{1}{3}(a^3+b^3+c^3)\left(\frac{1}{(1+b)(1+c)}+\frac{1}{(1+a)(1+c)}+\frac{1}{(1+a)(1+b)}\right)\\ &= \frac{1}{3}(a^3+b^3+c^3)\frac{3+(a+b+c)}{(1+a)(1+b)(1+c)}.\end{aligned}$$

Finally, use the facts that $\frac{1}{3}(a^3+b^3+c^3) \geq (\frac{a+b+c}{3})^3$, $\frac{a+b+c}{3} \geq 1$ and $(1+a)(1+b)(1+c) \leq \left(\frac{3+a+b+c}{3}\right)^3$ to see that

$$\frac{1}{3}(a^3+b^3+c^3)\frac{3+(a+b+a)}{(1+a)(1+b)(1+c)} \geq \left(\frac{a+b+c}{3}\right)^3 \frac{6}{(1+\frac{a+b+c}{3})^3} \geq \frac{6}{8}.$$

For the last inequality, notice that $\frac{\frac{a+b+c}{3}}{1+\frac{a+b+c}{3}} \geq \frac{1}{2}$.

Second solution. Multiplying by the common denominator and expanding both sides, the desired inequality becomes

$$4(a^4+b^4+c^4+a^3+b^3+c^3) \geq 3(1+a+b+c+ab+bc+ca+abc).$$

Since $4(a^4+b^4+c^4+a^3+b^3+c^3) = 4(3[4,0,0]+3[3,0,0])$ and $3(1+a+b+c+ab+bc+ca+abc) = 3([0,0,0]+3[1,0,0]+3[1,1,0]+[1,1,1])$, the inequality is equivalent to

$$4[4,0,0]+4[3,0,0] \geq [0,0,0]+3[1,0,0]+3[1,1,0]+[1,1,1].$$

Now, note that

$$[4,0,0] \geq \left[\frac{4}{3},\frac{4}{3},\frac{4}{3}\right] = a^{\frac{4}{3}}b^{\frac{4}{3}}c^{\frac{4}{3}} = 1 = [0,0,0],$$

where it has been used that $abc = 1$. Also,

$$3[4,0,0] \geq 3[2,1,1] = 3\frac{1}{3}(a^2bc + b^2ca + c^2ab) = 3\frac{1}{3}(a+b+c) = 3[1,0,0]$$

and

$$3[3,0,0] \geq 3\left[\frac{4}{3},\frac{4}{3},\frac{1}{3}\right] = 3\frac{1}{3}\left(a^{\frac{4}{3}}b^{\frac{4}{3}}c^{\frac{1}{3}} + b^{\frac{4}{3}}c^{\frac{4}{3}}a^{\frac{1}{3}} + c^{\frac{4}{3}}a^{\frac{4}{3}}b^{\frac{1}{3}}\right)$$
$$= 3\frac{1}{3}(ab+bc+ca) = 3[1,0,0].$$

Finally, $[3,0,0] \geq [1,1,1]$. Adding these results, we get the desired inequality.

4.2 Solutions to the exercises in Chapter 2

Solution 2.1. (i) Draw a segment BC of length a, a circle with radius c and center in B, and a circle with radius b and center in C, under what circumstances do they intersect?
(ii) It follows from (i).
(iii) $a = x+y$, $b = y+z$, $c = z+x \Leftrightarrow x = \frac{a+c-b}{2}$, $y = \frac{a+b-c}{2}$, $z = \frac{b+c-a}{2}$.

Solution 2.2. (i) $c < a+b \Rightarrow c < a+b+2\sqrt{ab} = (\sqrt{a}+\sqrt{b})^2 \Rightarrow \sqrt{c} < \sqrt{a}+\sqrt{b}$.
(ii) With 2, 3 and 4 it is possible to construct a triangle but with 4, 9 and 16 it is not possible to do so.
(iii) $a < b < c \Rightarrow a+b < a+c < b+c \Rightarrow \frac{1}{b+c} < \frac{1}{c+a} < \frac{1}{a+b}$, then it is sufficient to see that $\frac{1}{a+b} < \frac{1}{b+c} + \frac{1}{c+a}$, and it will be even easier to see that $\frac{1}{c} < \frac{1}{b+c} + \frac{1}{c+a}$.

Solution 2.3. Use the fact that if a, b, c are the lengths of the sides of a triangle, the angle that is opposed to the side c is either $90°$ or acute or obtuse if c^2 is equal, less or greater than a^2+b^2, respectively. Now, suppose that $a \leq b \leq c \leq d \leq e$ and that the segments (a,b,c) and (c,d,e) do not form an acute triangle; since $c^2 \geq a^2+b^2$ and $e^2 \geq c^2+d^2$, we deduce that $e^2 \geq a^2+b^2+d^2 \geq a^2+b^2+c^2 \geq a^2+b^2+a^2+b^2 = (a+b)^2+(a-b)^2 \geq (a+b)^2$, hence $a+b \leq e$, which is a contradiction.

Solution 2.4. Since $\angle A > \angle B$ then $BC > CA$. Using the triangle inequality we obtain $AB < BC + CA$, and by the previous statement, $AB < 2BC$.

Solution 2.5. (i) Let O be the intersection point of the diagonals AC and BD. Apply the triangle inequality to the triangles ABO and CDO. Adding the inequalities, we get $AB + CD < AC + BD$. On the other hand, by hypothesis we have that $AB + BD < AC + CD$. Adding these last two inequalities we get $AB < AC$.
(ii) Let DE be parallel to BC, then $\angle EDA < \angle BCD < \angle A$; therefore $DE > \frac{1}{2}AD$ and hence $\frac{1}{2}AD < DE < BC$. Refer to the previous exercise.

Solution 2.6. Each d_i is less than the sum of the lengths of two sides. Also, use the fact that in a convex quadrilateral the sum of the lengths of two opposite sides is less than the sum of the lengths of the diagonals.

Solution 2.7. Use the triangle inequality in the triangles ABA' and $AA'C$ to prove that $c < m_a + \frac{1}{2}a$ and $b < m_a + \frac{1}{2}a$.

Solution 2.8. If α, β, γ are the angles of a triangle in A, B and C, respectively, and if $\alpha_1 = \angle BAA'$ and $\alpha_2 = \angle A'AC$, then, using D2, $\beta > \alpha_1$ and $\gamma > \alpha_2$. Therefore, $180° = \alpha + \beta + \gamma > \alpha_1 + \alpha_2 + \alpha = 2\alpha$. Or, if we draw a circle with diameter BC, A should lie outside the circle and then $\angle BAC < 90°$.

Solution 2.9. Construct a parallelogram $ABDC$, with one diagonal BC and the other AD which is equal to two times the length of AA' and use D2 on the triangle ABD.

Solution 2.10. Complete a parallelogram as in the previous solution to prove that $m_a < \frac{b+c}{2}$. Similarly, $m_b < \frac{a+c}{2}$ and $m_c < \frac{a+b}{2}$. To prove the left hand side inequality, let A', B' and C' be the midpoints of the sides BC, CA and AB, respectively.

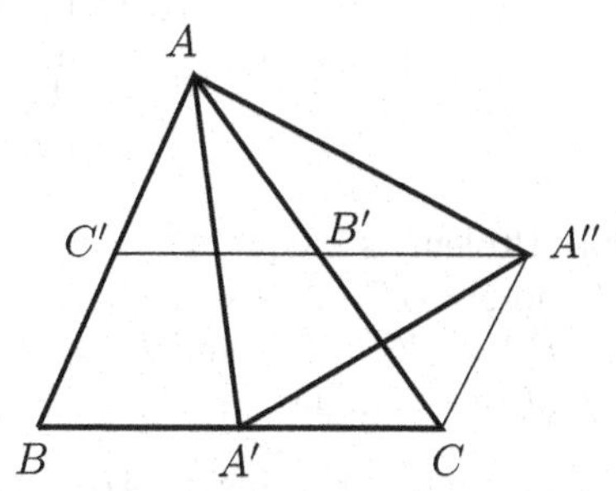

Extend the segment $C'B'$ to a point A'' such that $C'A'' = BC$. Apply the previous result to the triangle $AA'A''$ with side-lengths m_a, m_b and m_c.

Solution 2.11. Consider the quadrilateral $ABCD$ and let O be a point on the exterior of the quadrilateral so that AOB is similar to ACD, and thus OAC and BAD are also similar. If O, B and C are collinear, we have an equality, otherwise we have an inequality.[17]

Solution 2.12. Set $a = AB$, $b = BC$, $c = CD$, $d = DA$, $m = AC$ and $n = BD$. Let R be the radius of the circumcircle of $ABCD$. Thus we have[18]

$$(ABCD) = (ABC) + (CDA) = \frac{m(ab+cd)}{4R},$$

$$(ABCD) = (BCD) + (DAB) = \frac{n(bc+ad)}{4R}.$$

[17] See [6, page 136] or [1, page 128].
[18] See [6, page 97] or [9, page 13].

Therefore

$$\frac{m}{n} = \frac{bc+ad}{ab+cd} > 1 \Leftrightarrow bc+ad > ab+cd$$
$$\Leftrightarrow (d-b)(a-c) > 0.$$

Solution 2.13. Apply to the triangle ABP a rotation of $60°$ with center at A. Under the rotation the point B goes to the point C, and let P' be the image of P. The triangle $PP'C$ has as sides $PP' = PA$, $P'C = PB$ and PC, and then the result.

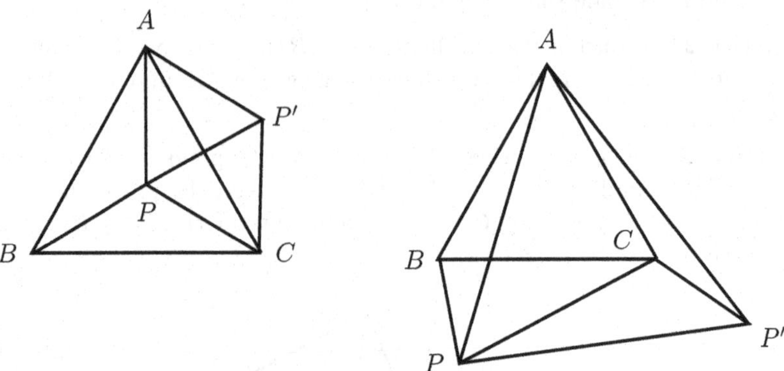

Second solution. Apply Ptolemy's inequality (see Exercise 2.11) to the quadrilaterals $ABCP$, $ABPC$ and $APBC$; after cancellation of common terms we obtain that $PB < PC + PA$, $PA < PC + PB$ and $PC < PA + PB$, respectively, which establish the existence of the triangle.

Third solution. For the case when P is inside ABC. Let P' be the point where AP intersects the side BC. Next, use that $AP < AP' < AB = BC < PB + PC$. In a similar way, the other inequalities $PB < PC + PA$ and $PC < PA + PB$ hold.

Solution 2.14. Set $a = AB$, $b = BC$, $x = AC$, $y = BD$. Remember that in a paralelogram we have $2(a^2 + b^2) = x^2 + y^2$. We can suppose, without loss of generality, that $a \leq b$. It is clear that $2b < x + y$, therefore $(2b)^2 < (x + y)^2 = x^2 + y^2 + 2xy = 2(a^2 + b^2) + 2xy$. Simplifying, we get $2(b^2 - a^2) < 2xy$.

Solution 2.15. (i) Extend the medians AA', BB' and CC' until they intersect the circumcircle at A_1, B_1 and C_1, respectively. Use the power of A' to establish that $A'A_1 = \frac{a^2}{4m_a}$. Also, use the facts that $m_a + A'A_1 \leq 2R$ and that the length of the median satisfies $m_a^2 = \frac{2(b^2+c^2)-a^2}{4}$, that is, $4m_a^2 + a^2 = 2(b^2 + c^2)$. We have analogous expressions for m_b and m_c.
(ii) Use Ptolemy's inequality in the quadrilaterals $AC'GB'$, $BA'GC'$ and $CB'GA'$, where G denotes the centroid. For instance, from the first quadrilateral we get $\frac{2}{3}m_a\frac{a}{2} \leq \frac{b}{2}\frac{m_c}{3} + \frac{c}{2}\frac{m_b}{3}$, then $2m_a a^2 \leq abm_c + cam_b$.

Solution 2.16. Using the formula $4m_b^2+b^2 = 2(c^2+a^2)$, we observe that $m_b^2-m_c^2 = \frac{3}{4}(c^2-b^2)$. Now, using the triangle inequality, prove that $m_b+m_c < \frac{3}{2}(b+c)$. From this you can deduce the left-hand side inequality.

The right-hand side inequality can be obtained from the first when applied to the triangle of sides[19] with lengths m_a, m_b and m_c.

Solution 2.17. Let a, b, c be the lengths of the sides of ABC. If E and F are the projections of I_a on the sides AB and CA, respectively, it is clear that if r_a is the radius of the excircle, we have that $r_a = I_aE = EA = AF = FI_a = s$, where s is the semiperimeter of ABC. Also, if h_a is the altitude of the triangle ABC from vertex A, then $\frac{AD}{DI_a} = \frac{h_a}{r_a}$. Since $ah_a = bc$, we have that

$$\frac{AD}{DI_a} = \frac{h_a}{r_a} = \frac{bc}{as} = \left(\frac{abc}{4R}\right)\left(\frac{4Rr}{a^2}\right)\left(\frac{1}{rs}\right) = \frac{4Rr}{a^2},$$

where r and R are the inradius and the circumradius of ABC, respectively. Since $2R = a$ and $2r = b+c-a$, therefore $\frac{AD}{DI_a} = \frac{b+c-a}{a} = \frac{b+c}{a} - 1$. Then, it is enough to prove that $\frac{b+c}{a} \leq \sqrt{2}$ or, equivalently, that $2bc \leq a^2$, but $bc = \sqrt{b^2c^2} \leq \frac{b^2+c^2}{2} = \frac{a^2}{2}$.

Solution 2.18. Simplifying, the first inequality is equivalent to $ab + bc + ca \leq a^2+b^2+c^2$, which follows from Exercise 1.27. For the second one, expand $(a+b+c)^2$ and use the triangle inequality to obtain $a^2 < a(b+c)$.

Solution 2.19. Use the previous suggestion.

Solution 2.20. Expand and you will get the previous exercise.

Solution 2.21. The first inequality is the Nesbitt's inequality, Example 1.4.8. For the second inequality use the fact that $a+b > \frac{a+b+c}{2}$, then $\frac{c}{a+b} < \frac{2c}{a+b+c}$.

Solution 2.22. Observe that $a^2(b+c-a)+b^2(c+a-b)+c^2(a+b-c)-2abc = (b+c-a)(c+a-b)(a+b-c)$, now see Example 2.2.3.

Solution 2.23. Observe that

$$\begin{aligned} a\left(b^2+c^2-a^2\right) &+ b\left(c^2+a^2-b^2\right)+c\left(a^2+b^2-c^2\right) \\ &= a^2(b+c-a)+b^2(c+a-b)+c^2(a+b-c), \end{aligned}$$

now see Exercise 2.22.

Solution 2.24. Use Ravi's transformation with $a = y+z$, $b = z+x$, $c = x+y$ to see first that

$$a^2b(a-b)+b^2c(b-c)+c^2a(c-a) = 2(xy^3+yz^3+zx^3) - 2(xy^2z+x^2yz+xyz^2).$$

Then, the inequality is equivalent to $\frac{x^2}{y}+\frac{y^2}{z}+\frac{z^2}{x} \geq x+y+z$. Apply then inequality (1.11).

[19] See the solution of Exercise 2.10.

Solution 2.25.

$$\left|\frac{a-b}{a+b}+\frac{b-c}{b+c}+\frac{c-a}{c+a}\right| = \left|\frac{a-b}{a+b}\cdot\frac{b-c}{b+c}\cdot\frac{c-a}{c+a}\right|$$
$$< \frac{cab}{(a+b)(b+c)(c+a)} \le \frac{1}{8}.$$

For the last inequality, see the solution of Example 2.2.3.

Solution 2.26. By Exercise 2.18,

$$3(ab+bc+ca) \le (a+b+c)^2 \le 4(ab+bc+ca).$$

Then, since $ab+bc+ca=3$, it follows that $9 \le (a+b+c)^2 \le 12$, and then the result.

Solution 2.27. Use Ravi's transformation, $a=y+z$, $b=z+x$ and $c=x+y$. The AM-GM inequality and the Cauchy-Schwarz inequality imply

$$\begin{aligned}\frac{1}{a}+\frac{1}{b}+\frac{1}{c} &= \frac{1}{y+z}+\frac{1}{z+x}+\frac{1}{x+y}\\ &\le \frac{1}{2}\left(\frac{1}{\sqrt{yz}}+\frac{1}{\sqrt{zx}}+\frac{1}{\sqrt{xy}}\right)\\ &= \frac{\sqrt{x}+\sqrt{y}+\sqrt{z}}{2\sqrt{xyz}}\\ &\le \frac{\sqrt{3}\sqrt{x+y+z}}{2\sqrt{xyz}}\\ &= \frac{\sqrt{3}}{2}\sqrt{\frac{x+y+z}{xyz}} = \frac{\sqrt{3}}{2r}.\end{aligned}$$

For the last identity, see the end of the proof of Example 2.2.4.

Solution 2.28. The part (i) follows from the following equivalences:

$$\begin{aligned}(s-a)(s-b) < ab &\Leftrightarrow s^2 - s(a+b) < 0\\ &\Leftrightarrow a+b+c < 2(a+b)\\ &\Leftrightarrow c < a+b.\end{aligned}$$

For (ii), use Ravi's transformation, $a=y+z$, $b=z+x$, $c=x+y$, in order to see that the inequality is equivalent to

$$4(xy+yz+zx) \le (y+z)(z+x)+(z+x)(x+y)+(x+y)(y+z).$$

In turn, the last inequality follows from the inequality $xy+yz+zx \le x^2+y^2+z^2$, which is Exercise 1.27.

Another way to obtain (ii) is the following: the given inequality is equivalent to $3s^2 - 2s(a+b+c) + (ab+bc+ca) \leq \frac{ab+bc+ca}{4}$, which in turn is equivalent to $3(ab+bc+ca) \leq 4s^2$. The last inequality can be rewritten as $3(ab+bc+ca) \leq (a+b+c)^2$.

Solution 2.29. Applying the cosine law, we can see that

$$\begin{aligned}\sqrt{a^2+b^2-c^2}\sqrt{a^2-b^2+c^2} &= \sqrt{2ab\cos C}\sqrt{2ac\cos B}\\ &= 2a\sqrt{(b\cos C)(c\cos B)}\\ &\leq 2a\frac{b\cos C + c\cos B}{2} = a^2.\end{aligned}$$

Solution 2.30. Using the Cauchy-Schwarz inequality, for any x, y, z, $w \geq 0$, we have that

$$\sqrt{xy} + \sqrt{zw} \leq \sqrt{(x+z)(y+w)}.$$

Therefore

$$\begin{aligned}\sum_{\text{cyclic}} \sqrt{a^2+b^2-c^2}\sqrt{a^2-b^2+c^2} &= \frac{1}{2}\sum_{\text{cyclic}}\Big(\sqrt{a^2+b^2-c^2}\sqrt{a^2-b^2+c^2}\\ &\quad + \sqrt{c^2+a^2-b^2}\sqrt{c^2-a^2+b^2}\Big)\\ &\leq \frac{1}{2}\sum_{\text{cyclic}}\sqrt{(2a^2)(2c^2)} = \sum_{\text{cyclic}} ac.\end{aligned}$$

Solution 2.31. Consider positive numbers x, y, z with $a = y+z$, $b = z+x$ and $c = x+y$. The inequalities are equivalent to proving that

$$\frac{y+z}{2x} + \frac{z+x}{2y} + \frac{x+y}{2z} \geq 3 \quad \text{and} \quad \frac{2x}{y+z} + \frac{2y}{z+x} + \frac{2z}{x+y} \geq 3.$$

For the first inequality use the fact that $\frac{y}{x} + \frac{x}{y} \geq 2$ and for the second inequality use Nesbitt's inequality.

Solution 2.32. Since in triangles with the same base, the ratio between its altitudes is equal to the ratio of theirs areas, we have that

$$\frac{PQ}{AD} + \frac{PR}{BE} + \frac{PS}{CF} = \frac{(PBC)}{(ABC)} + \frac{(PCA)}{(ABC)} + \frac{(PAB)}{(ABC)} = \frac{(ABC)}{(ABC)} = 1.$$

Use inequality (2.3) of Section 2.3.

Solution 2.33. (i) Recall that $(S_1+S_2+S_3)(\frac{1}{S_1}+\frac{1}{S_2}+\frac{1}{S_3}) \geq 9$.

(ii) The non-common vertices of the triangles form a hexagon which is divided into 6 triangles S_1, S_2, S_3, T_1, T_2, T_3, where S_i and T_i have one common angle.

Using the formula for the area that is related to the sine of the angle, prove that $S_1S_2S_3 = T_1T_2T_3$. After this, use the AM-GM inequality as follows:

$$\begin{aligned} S\left(\frac{1}{S_1}+\frac{1}{S_2}+\frac{1}{S_3}\right) &\geq (S_1+S_2+S_3+T_1+T_2+T_3)\left(\frac{1}{S_1}+\frac{1}{S_2}+\frac{1}{S_3}\right) \\ &\geq \frac{18\sqrt[6]{S_1S_2S_3T_1T_2T_3}}{\sqrt[3]{S_1S_2S_3}} = 18. \end{aligned}$$

The equality holds when the point O is the centroid of the triangle and the lines through O are the medians of the triangle; in this case $S_1 = S_2 = S_3 = T_1 = T_2 = T_3 = \frac{1}{6}S$.

Solution 2.34. If $P = G$ is the centroid, the equality is evident since $\frac{AG}{GL} = \frac{BG}{GM} = \frac{CG}{GN} = 2$.

On the other hand, if $\frac{AP}{PL} + \frac{BP}{PM} + \frac{CP}{PN} = 6$, we have $\frac{AL}{PL} + \frac{BM}{PM} + \frac{CN}{PN} = 9$. It is not difficult to see that $\frac{PL}{AL} = \frac{(PBC)}{(ABC)}$, $\frac{PM}{BM} = \frac{(PCA)}{(ABC)}$ and $\frac{PN}{CN} = \frac{(PAB)}{(ABC)}$, therefore $\frac{PL}{AL} + \frac{PM}{BM} + \frac{PN}{CN} = 1$. This implies that

$$\left(\frac{AL}{PL}+\frac{BM}{PM}+\frac{CN}{PN}\right)\left(\frac{PL}{AL}+\frac{PM}{BM}+\frac{PN}{CN}\right) = 9.$$

By inequality (2.3), the equality above holds only in the case when $\frac{AL}{PL} = \frac{BM}{PM} = \frac{CN}{PN} = 3$, which implies that P is the centroid.

Solution 2.35. (i) It is known that $HD = DD'$, $HE = EE'$ and $HF = FF'$, where H is the orthocenter.[20] Thus, the solution follows from part (i) of Example 2.3.4. (ii) Since $\frac{AD'}{AD} = \frac{AD+DD'}{AD} = 1 + \frac{HD}{AD}$, we also have, after looking at the solution to Example 2.3.4, that $\frac{AD'}{AD} + \frac{BE'}{BE} + \frac{CF'}{CF} = 1 + \frac{HD}{AD} + 1 + \frac{HE}{BE} + 1 + \frac{HF}{CF} = 4$.

Since $\left(\frac{AD}{AD'} + \frac{BE}{BE'} + \frac{CF}{CF'}\right)\left(\frac{AD'}{AD} + \frac{BE'}{BE} + \frac{CF'}{CF}\right) \geq 9$, we have the result.

Solution 2.36. As it has been mentioned in the proof of Example 2.3.5, the length of the internal bisector of angle A satisfies

$$l_a^2 = bc\left(1-\left(\frac{a}{b+c}\right)^2\right) = \frac{4bc}{(b+c)^2}(s(s-a)).$$

Since $4bc \leq (b+c)^2$, it follows that $l_a^2 \leq s(s-a)$ and $l_al_b \leq s\sqrt{(s-a)(s-b)} \leq s\frac{(s-a)+(s-b)}{2} = s\frac{c}{2}$.

Therefore, $l_al_bl_c \leq s\sqrt{s(s-a)(s-b)(s-c)} = s(sr)$, $l_al_b + l_bl_c + l_cl_a \leq s\left(\frac{a+b+c}{2}\right) = s^2$ and $l_a^2 + l_b^2 + l_c^2 \leq s(s-a) + s(s-b) + s(s-c) = s^2$.

[20] See [6, page 85] or [9, page 37].

Solution 2.37. Let $\alpha = \angle AMB$, $\beta = \angle BNA$, $\gamma = \angle APC$, and let (ABC) be the area of ABC. We have

$$(ABC) = \frac{1}{2}a \cdot AM \sin\alpha = \frac{abc}{4R}.$$

Hence, $\frac{bc}{AM} = 2R\sin\alpha$. Similarly, $\frac{ca}{BN} = 2R\sin\beta$ and $\frac{ab}{CP} = 2R\sin\gamma$. Thus,

$$\frac{bc}{AM} + \frac{ca}{BN} + \frac{ab}{CP} = 2R(\sin\alpha + \sin\beta + \sin\gamma) \le 6R.$$

Equality is attained if M, N and P are the feet of the altitudes.

Solution 2.38. Let A_1, B_1, C_1 be the midpoints of the sides BC, CA, AB, respectively, and let B_2, C_2 be the reflections of A_1 with respect to AB and CA, respectively. Also, consider D as the intersection of AB with A_1B_2 and E the intersection of CA with A_1C_2. Then,

$$2DE = B_2C_2 \le C_2B_1 + B_1C_1 + C_1B_2 = A_1B_1 + B_1C_1 + C_1A_1 = s.$$

Use the fact that the quadrilateral A_1DAE is inscribed on a circle of diameter AA_1 and the sine law on ADE, to deduce that $DE = AA_1 \sin A = m_a \sin A$. Then, $s \ge 2DE = 2m_a \sin A = 2m_a\frac{a}{2R} = \frac{am_a}{R}$, that is, $am_a \le sR$. Similarly, we have that $bm_b \le sR$ and $cm_c \le sR$.

Solution 2.39. The inequality is equivalent to $8(s-a)(s-b)(s-c) \le abc$, where s is the semiperimeter.

Since $(ABC) = sr = \frac{abc}{4R} = \sqrt{s(s-a)(s-b)(s-c)}$, where r and R denote the inradius and the circumradius of ABC, respectively; we only have to prove that $8sr^2 \le abc$, that is, $8sr^2 \le 4Rrs$, which is equivalent to $2r \le R$.

Solution 2.40. The area of a triangle ABC satisfies the equalities $(ABC) = \frac{abc}{4R} = \frac{(a+b+c)r}{2}$, therefore $\frac{1}{ab}+\frac{1}{bc}+\frac{1}{ca} = \frac{1}{2Rr} \ge \frac{1}{R^2}$, where R and r denote the circumradius and the inradius, respectively.

Solution 2.41. Use Exercise 2.40 and the sine law.

Solution 2.42. Use that[21] $\sin\frac{A}{2} = \sqrt{\frac{(s-b)(s-c)}{bc}}$, where s denotes the semiperimeter of the triangle ABC, and similar expressions for $\sin\frac{B}{2}$ and $\sin\frac{C}{2}$, to see that

$$\sin\frac{A}{2}\sin\frac{B}{2}\sin\frac{C}{2} = \frac{(s-a)(s-b)(s-c)}{abc} = \frac{sr^2}{abc} = \frac{r}{4R} \le \frac{1}{8},$$

where R and r are the circumradius and the inradius of ABC, respectively.

[21] Notice that $\sin^2\frac{A}{2} = \frac{1-\cos A}{2} = \frac{1-\frac{b^2+c^2-a^2}{2bc}}{2} = \frac{a^2-(b-c)^2}{4bc} = \frac{(s-b)(s-c)}{bc}$.

Solution 2.43. From inequality (2.3), we know that

$$(a+b+c)\left(\frac{1}{a}+\frac{1}{b}+\frac{1}{c}\right) \geq 9.$$

Since $a+b+c \leq 3\sqrt{3}R$, we have

$$\frac{1}{a}+\frac{1}{b}+\frac{1}{c} \geq \frac{\sqrt{3}}{R}. \tag{4.4}$$

Applying, once more, inequality (2.3), we get

$$\frac{1}{3}\left(\frac{\pi}{2A}+\frac{\pi}{2B}+\frac{\pi}{2C}\right) \geq \frac{3}{\frac{2}{\pi}(A+B+C)} = \frac{3}{2}. \tag{4.5}$$

Let $f(x) = \log \frac{\pi}{2x}$, since $f''(x) = \frac{1}{x^2} > 0$, f is convex. Using Jensen's inequality, we get

$$\frac{1}{3}\left(\log\frac{\pi}{2A}+\log\frac{\pi}{2B}+\log\frac{\pi}{2C}\right) \geq \log\left[\frac{1}{3}\left(\frac{\pi}{2A}+\frac{\pi}{2B}+\frac{\pi}{2C}\right)\right].$$

Applying (4.5) and the fact that $\log x$ is a strictly increasing function, we obtain

$$\frac{1}{3}\left(\log\frac{\pi}{2A}+\log\frac{\pi}{2B}+\log\frac{\pi}{2C}\right) \geq \log\frac{3}{2}. \tag{4.6}$$

We can suppose that $a \leq b \leq c$, which implies $A \leq B \leq C$. Therefore $\frac{1}{a} \geq \frac{1}{b} \geq \frac{1}{c}$ and $\log\frac{\pi}{2A} \geq \log\frac{\pi}{2B} \geq \log\frac{\pi}{2C}$. Using Tchebyshev's inequality,

$$\frac{1}{a}\log\frac{\pi}{2A}+\frac{1}{b}\log\frac{\pi}{2B}+\frac{1}{c}\log\frac{\pi}{2C} \geq \left(\frac{1}{a}+\frac{1}{b}+\frac{1}{c}\right)\left(\frac{\log\frac{\pi}{2A}+\log\frac{\pi}{2B}+\log\frac{\pi}{2C}}{3}\right).$$

Therefore, using (4.4) and (4.6) leads us to

$$\frac{1}{a}\log\frac{\pi}{2A}+\frac{1}{b}\log\frac{\pi}{2B}+\frac{1}{c}\log\frac{\pi}{2C} \geq \frac{\sqrt{3}}{R}\log\frac{3}{2}.$$

Now, raising the expresions to the appropriate powers and taking the reciprocals, we obtain the desired inequality. In all the above inequalities, the equality holds if and only if $a = b = c$ (this means, equality is obtained if and only if the triangle is equilateral).

Solution 2.44. By the sine law, it follows that

$$\frac{\sin A}{a} = \frac{\sin B}{b} = \frac{\sin C}{c} = \frac{1}{2R},$$

where a, b, c are the lengths of the sides of the triangle and R is the circumradius of the triangle. Thus,

$$\begin{aligned}\sin^2 A + \sin^2 B + \sin^2 C &= \frac{a^2}{4R^2} + \frac{b^2}{4R^2} + \frac{c^2}{4R^2}\\ &= \frac{1}{4R^2}(a^2+b^2+c^2)\\ &\le \frac{1}{4R^2}\cdot 9R^2 = \frac{9}{4},\end{aligned}$$

where the inequality follows from Leibniz's inequality.

Solution 2.45. Use Leibniz's inequality and the fact that the area of a triangle is given by $(ABC) = \frac{abc}{4R}$.

Solution 2.46. We note that the incircle of ABC is the circumcircle of DEF. Applying Leibniz's inequality to DEF, we get

$$EF^2 + FD^2 + DE^2 \le 9r^2,$$

where r is the inradius of ABC. On the other hand, using Theorem 2.4.3 we obtain $s^2 \ge 27r^2$, hence

$$EF^2 + FD^2 + DE^2 \le \frac{s^2}{3}.$$

Solution 2.47.

$$\begin{aligned}\frac{a^2}{h_b h_c} + \frac{b^2}{h_c h_a} + \frac{c^2}{h_a h_b} &= \frac{a^2bc + b^2ca + c^2ab}{4(ABC)^2} = \frac{abc(a+b+c)}{4(ABC)^2}\\ &= \frac{abc(a+b+c)}{4\frac{abc}{4R}\frac{(a+b+c)r}{2}} = \frac{2R}{r} \ge 4.\end{aligned}$$

Solution 2.48. Remember that $\sin^2\frac{A}{2} = \frac{1-\cos A}{2}$ and use that $\cos A + \cos B + \cos C \le \frac{3}{2}$ (see Example 2.5.2).

Solution 2.49. Observe that

$$4\sqrt{3}(ABC) \le \frac{9abc}{a+b+c} \Leftrightarrow 4\sqrt{3}rs \le \frac{9\cdot 4\,Rrs}{2s} \Leftrightarrow 2\sqrt{3}s \le 9\,R \Leftrightarrow \frac{2s}{3\sqrt{3}} \le R.$$

The last inequality was proved in Theorem 2.4.3.

Solution 2.50. Use the previous exercise and the inequality between the harmonic mean and the geometric mean,

$$\frac{3}{\frac{1}{ab}+\frac{1}{bc}+\frac{1}{ca}} \le \sqrt[3]{a^2b^2c^2}.$$

Solution 2.51. Use the previous exercise and the AM-GM inequality,

$$\sqrt[3]{a^2b^2c^2} \leq \frac{a^2+b^2+c^2}{3}.$$

Solution 2.52. First, observe that if $s = \frac{a+b+c}{2}$, then

$$\begin{aligned} a^2+b^2+c^2-(a-b)^2-(b-c)^2-(c-a)^2 &= \\ &= a^2-(b-c)^2+b^2-(c-a)^2+c^2-(a-b)^2 \\ &= 4\{(s-b)(s-c)+(s-c)(s-a)+(s-a)(s-b)\}. \end{aligned}$$

Hence, if $x = s-a$, $y = s-b$, $z = s-c$, then the inequality is equivalent to

$$\sqrt{3}\sqrt{xyz(x+y+z)} \leq xy+yz+zx.$$

Squaring and simplifying the last inequality, we get

$$xyz(x+y+z) \leq x^2y^2+y^2z^2+z^2x^2.$$

This inequality can be deduced using Cauchy-Schwarz's inequality with (xy, yz, zx) and (zx, xy, yz).

Solution 2.53. Use Exercise 2.50 and the inequality $3\sqrt[3]{(ab)(bc)(ca)} \leq ab+bc+ca$.

Solution 2.54. Note that

$$\begin{aligned} \frac{3(a+b+c)abc}{ab+bc+ca} \geq \frac{9abc}{a+b+c} \quad &\Leftrightarrow \quad (a+b+c)^2 \geq 3(ab+bc+ca) \\ &\Leftrightarrow \quad a^2+b^2+c^2 \geq ab+bc+ca, \end{aligned}$$

now, use Exercise 2.49.

Solution 2.55. Using (2.5), (2.6) and (2.7) we can observe that $a^2+b^2+c^2+4abc = \frac{1}{2} - 2r^2$.

Solution 2.56. Observe the relationships used in the proof of Exercise 2.39,

$$\begin{aligned} \frac{(b+c-a)(c+a-b)(a+b-c)}{abc} &= \frac{8(s-a)(s-b)(s-c)}{abc} \\ &= \frac{8s(s-a)(s-b)(s-c)}{4Rs(\frac{abc}{4R})} \\ &= \frac{8(rs)^2}{4Rs(rs)} = \frac{2r}{R}. \end{aligned}$$

Solution 2.57. Observe that

$$
\begin{aligned}
\frac{a^2}{b+c-a}+\frac{b^2}{c+a-b}+\frac{c^2}{a+b-c} &= \frac{1}{2}\left(\frac{a^2}{s-a}+\frac{b^2}{s-b}+\frac{c^2}{s-c}\right)\\
&= \frac{1}{2}\left(\frac{sa}{s-a}-a+\frac{sb}{s-b}-b+\frac{sc}{s-c}-c\right)\\
&= \frac{s}{2}\left(\frac{a}{s-a}+\frac{b}{s-b}+\frac{c}{s-c}\right)-s\\
&= \frac{s}{2}\left[\frac{(a+b+c)s^2-2(ab+bc+ca)s+3abc}{(s-a)(s-b)(s-c)}\right]-s\\
&= \frac{s}{2}\left[\frac{2s^3-2s(s^2+r^2+4rR)+3(4Rrs)}{r^2s}\right]-s\\
&= \frac{2s(R-r)}{r} \geq \frac{2s(R-\frac{R}{2})}{r} \geq \frac{3\sqrt{3}rR}{r} = 3\sqrt{3}R,
\end{aligned}
$$

the last two inequalities follow from the fact that $R \geq 2r$ (which implies that $-r \geq \frac{-R}{2}$) and from $s \geq 3\sqrt{3}r$, respectively.

Solution 2.58. Start on the side of the equations which expresses the relationship between the τ's and perform the operations.

Solution 2.59. If x_1, $1-x_1$, x_2, $1-x_2$, ... are the lengths into which each side is divided for the corresponding point, we can deduce that $a^2+b^2+c^2+d^2 = \sum(x_i^2+(1-x_i)^2)$. Prove that $\frac{1}{2} \leq 2(x_i-\frac{1}{2})^2+\frac{1}{2} = x_i^2+(1-x_i)^2 \leq 1$.

For part (ii), the inequality on the right-hand side follows from the triangle inequality. For the one on the left-hand side, use reflections on the sides, as you can see in the figure.

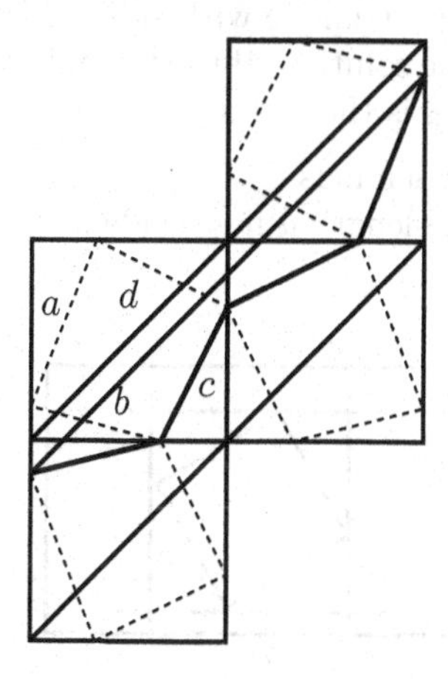

Solution 2.60. This is similar to part (ii) of the previous problem.

Solution 2.61. If ABC is the triangle and $DEFGHI$ is the hexagon with DE, FG, HI parallel to BC, AB, CA, respectively, we have that the perimeter of the hexagon is $2(DE+FG+HI)$. Let X, Y, Z be the tangency points of the incircle with the sides BC, CA, AB, respectively, and let $p=a+b+c$ be the perimeter of the triangle ABC. Set $x=AZ=AY$, $y=BZ=BX$ and $z=CX=CY$, then we have the relations

$$\frac{DE}{a}=\frac{AE+ED+DA}{p}=\frac{2x}{p}.$$

Similarly, we have the other relations

$$\frac{FG}{c}=\frac{2z}{p},\quad \frac{HI}{b}=\frac{2y}{p}.$$

Therefore,

$$\begin{aligned} p(DEFGHI)=\frac{4(xa+yb+zc)}{p} &= \frac{4(a(s-a)+b(s-b)+c(s-c))}{2s} \\ &= \frac{4((a+b+c)s-(a^2+b^2+c^2))}{2s} \\ &= 2(a+b+c)-4\frac{(a^2+b^2+c^2)}{(a+b+c)}, \end{aligned}$$

but $a^2+b^2+c^2\geq\frac{1}{3}(a+b+c)(a+b+c)$ by Tchebyshev's inequality. Thus, $p(DEFGHI)\leq 2(a+b+c)-\frac{4}{3}(a+b+c)=\frac{2}{3}(a+b+c)$.

Solution 2.62. Take the circumcircle of the equilateral triangle with side length 2. The circles with centers the midpoints of the sides of the triangle and radii 1 cover a circle of radius 2. If a circle of radius greater than $\frac{2\sqrt{3}}{3}$ is covered by three circles of radius 1, then one of the three circles covers a chord of length greater than 2.

Solution 2.63. Take the acute triangle with sides of lengths $2r_1$, $2r_2$ and $2r_3$, if it exists. Its circumradius is the solution. If the triangle does not exist, the maximum radius between r_1, r_2 and r_3 is the answer.

Solution 2.64. We need two lemmas.
Lemma 1. If a square of side-length a lies inside a rectangle of sides c and d, then $a\leq\min\{c,d\}$.

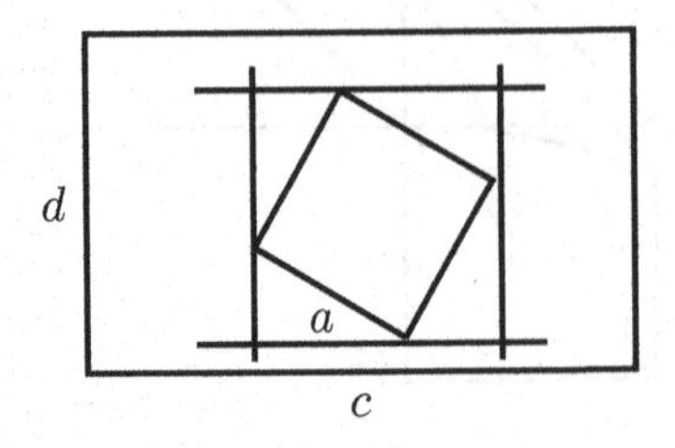

Through the vertices of the square draw parallel lines to the sides of the rectangle in such a way that those lines enclose the square as in the figure. Since the parallel lines form a square inside the rectangle and such a square contains the original square, we have the result.

Lemma 2. The diagonal of a square inscribed in a right triangle is less than or equal to the length of the internal bisector of the right angle.

Let ABC be the right triangle with hypotenuse CA and let $PQRS$ be the inscribed square.

It can be assumed that the vertices P and Q belong to the legs of the right triangle (otherwise, translate the square) and let O be the intersection point of the diagonals PR and QS.

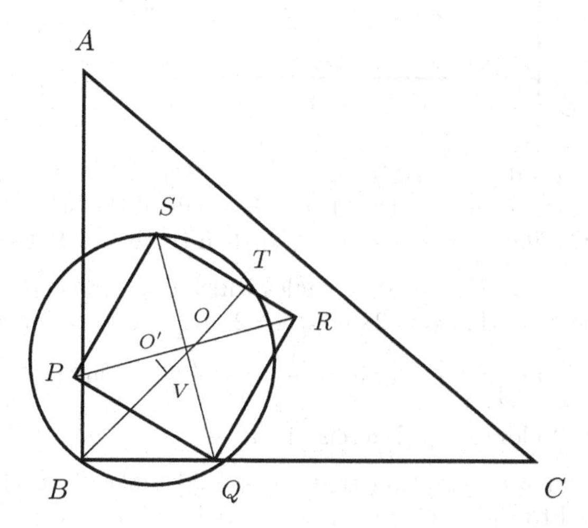

Since $BQOP$ is cyclic ($\angle B = \angle O = 90°$), it follows that $\angle QBO = \angle QPO = 45°$, then O belongs to the internal bisector of $\angle B$. Let T be the intersection of BO with RS, then $\angle QBT = \angle QST = 45°$, therefore $BQTS$ is cyclic and the center O' of the circumcircle of $BQTS$ is the intersection of the perpendicular bisectors of SQ and BT. But the perpendicular bisector of SQ is PR, hence the point O' belongs to PR, and if V is the midpoint of BT, we have that VOO' is a right triangle. Since $O'O > O'V$, then the chords SQ and BT satisfy $SQ < BT$, and the lemma follows.

Let us finish now the proof of the problem. Let $ABCD$ be the square of side 1 and let l be a line that separates the two squares. If l is parallel to one of the sides of the square $ABCD$, then Lemma 1 applies. Otherwise, l intersects every line that determines a side of the square $ABCD$. Suppose that A is the farthest vertex from l.

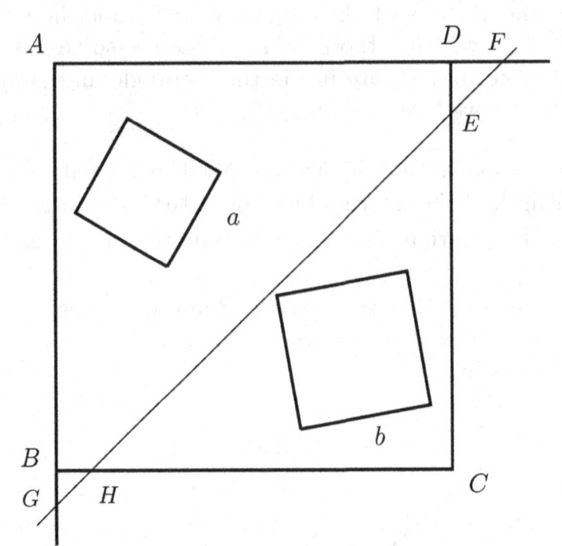

If l intersects the sides of $ABCD$ in E, F, G, H as in the figure, we have, by Lemma 2, that the sum of the lengths of the diagonals of the small squares is less than or equal to AC, that is, $\sqrt{2}(a+b) \leq \sqrt{2}$, then the result follows.

Solution 2.65. If α, β, γ are the central angles which open the chords of length a, b, c, respectively, we have that $a = 2\sin\frac{\alpha}{2}$, $b = 2\sin\frac{\beta}{2}$ and $c = 2\sin\frac{\gamma}{2}$. Therefore,

$$abc = 8\sin\frac{\alpha}{2}\sin\frac{\beta}{2}\sin\frac{\gamma}{2} \leq 8\sin^3\left(\frac{\alpha+\beta+\gamma}{6}\right) = 8\sin^3(30^\circ) = 1,$$

where the inequality follows from Exercise 1.81.

Solution 2.66. The first observation that we should make is to check that the diagonals are parallel to the sides. Let X be the point of intersection between the diagonals AD and CE. Now, the pentagon can be divided into

$$(ABCDE) = (ABC) + (ACX) + (CDE) + (EAX).$$

Since $ABCX$ is a parallelogram, we have $(ABC) = (CXA) = (CDE)$. Let $a = (CDX) = (EAX)$ and $b = (DEX)$, then we get $\frac{a}{b} = \frac{AX}{XD} = \frac{(CXA)}{(CDX)} = \frac{a+b}{a}$, that is, $\frac{a}{b} = \frac{1+\sqrt{5}}{2}$. Now we have all the elements to find $(ABCDE)$.

Solution 2.67. Prove that $sr = s_1R = (ABC)$, where s_1 is the semiperimeter of the triangle DEF. To deduce this equality, it is sufficient to observe that the radii OA, OB and OC are perpendicular to EF, FD and DE, respectively. Use also that $R \geq 2r$.

Solution 2.68. Suppose that the maximum angle is A and that it satisfies $60^\circ \leq A \leq 90^\circ$, then the lengths of the altitudes h_b and h_c are also less than 1. Now, use the fact that $(ABC) = \frac{h_b h_c}{2\sin A}$ and that $\frac{\sqrt{3}}{2} \leq \sin A \leq 1$. The obtuse triangle case is easier.

Solution 2.69. Let $ABCD$ be the quadrilateral with sides of length $a = AB$, $b = BC$, $c = CD$ and $d = DA$.
(i) $(ABCD) = (ABC) + (CDA) = \frac{ab\sin B}{2} + \frac{cd\sin D}{2} \leq \frac{ab+cd}{2}$.
(ii) If $ABCD$ is the quadrilateral mentioned with sides of length a, b, c and d, consider the triangle $BC'D$ which results from the reflection of DCB with respect to the perpendicular bisector of side BD. The quadrilaterals $ABCD$ and $ABC'D$ have the same area but the second one has sides of length a, c, b and d, in this order. Now use (i).
(iii) $(ABC) \leq \frac{ab}{2}$, $(BCD) \leq \frac{bc}{2}$, $(CDA) \leq \frac{cd}{2}$ and $(DAB) \leq \frac{da}{2}$.

Solution 2.70. In Example 2.7.6 we proved that

$$PA \cdot PB \cdot PC \geq \frac{R}{2r}(p_a + p_b)(p_b + p_c)(p_c + p_a).$$

Use the AM-GM inequality.

Solution 2.71. (i) $\frac{PA^2}{p_bp_c} + \frac{PB^2}{p_cp_a} + \frac{PC^2}{p_ap_b} \geq 3\sqrt[3]{\frac{PA^2}{p_bp_c}\frac{PB^2}{p_cp_a}\frac{PC^2}{p_ap_b}} \geq 3\sqrt[3]{\left(\frac{4R}{r}\right)^2} \geq 12.$
(ii) $\frac{PA}{p_b+p_c} + \frac{PB}{p_c+p_a} + \frac{PC}{p_a+p_b} \geq 3\sqrt[3]{\frac{PA}{p_b+p_c}\frac{PB}{p_c+p_a}\frac{PC}{p_a+p_b}} \geq 3\sqrt[3]{\frac{R}{2r}} \geq 3.$
(iii) $\frac{PA}{\sqrt{p_bp_c}} + \frac{PB}{\sqrt{p_cp_a}} + \frac{PC}{\sqrt{p_ap_b}} \geq 3\sqrt[3]{\frac{PA}{\sqrt{p_bp_c}}\frac{PB}{\sqrt{p_cp_a}}\frac{PC}{\sqrt{p_ap_b}}} \geq 3\sqrt[3]{\frac{4R}{r}} \geq 6.$

For the last inequalities in (i) and (iii), we have used Exercise 2.70. For the last inequality in (ii), we have resorted to Example 2.7.6.
(iv) Proceed as in Example 2.7.5, that is, apply inversion in a circle with center P and radius d (arbitrary, for instance $d = p_b$). Let A', B', C' be the inverses of A, B, C, respectively. Let p'_a, p'_b, p'_c be the distances from P to the sides $B'C'$, $C'A'$, $A'B'$, respectively.

Let us prove that $p'_a = \frac{p_aPB' \cdot PC'}{d^2}$. We have

$$p'_aB'C' = 2(PB'C') = \frac{PB' \cdot PC' \cdot B'C'}{PA'_1} = \frac{p_aPB' \cdot PC' \cdot B'C'}{d^2},$$

where A'_1 is the inverse of A_1, the projection of P on BC. Similarly, $p'_b = \frac{p_bPC' \cdot PA'}{d^2}$ and $p'_c = \frac{p_cPA' \cdot PB'}{d^2}$.

The Erdős-Mordell inequality, applied to the triangle $A'B'C'$, guarantees us that $PA' + PB' + PC' \geq 2(p'_a + p'_b + p'_c)$.

Now, since $PA \cdot PA' = PB \cdot PB' = PC \cdot PC' = d^2$, after substitution we get

$$\frac{1}{PA} + \frac{1}{PB} + \frac{1}{PC} \geq 2\left(\frac{p_a}{PB \cdot PC} + \frac{p_b}{PC \cdot PA} + \frac{p_c}{PC \cdot PA}\right)$$

and this inequality is equivalent to

$$PB \cdot PC + PC \cdot PA + PA \cdot PB \geq 2(p_aPA + p_bPB + p_cPC).$$

Finally, to conclude use example 2.7.4.

Solution 2.72. If P is an interior point or a point on the perimeter of the triangle ABC, see the proof of Theorem 2.7.2.

If h_a is the length of the altitude from vertex A, we have that the area of the triangle ABC satisfies $2(ABC) = ah_a = ap_a + bp_b + cp_c$.

Since $h_a \leq PA + p_a$ (even if $p_a \leq 0$, that is, if P is a point on the outside of the triangle, on a different side of BC than A), and because the equality holds if P is exactly on the segment of the altitude from the vertex A, therefore $aPA + ap_a \geq ah_a = ap_a + bp_b + cp_c$, hence $aPA \geq bp_b + cp_c$.

This inequality can be applied to triangle $AB'C'$ symmetric to ABC with respect to the internal angle bisector of A, where $aPA \geq cp_b + bp_c$, with equality when AP passes through the point O.

Similarly, $bPB \geq ap_c + cp_a$ and $cPC \geq ap_b + bp_a$, therefore

$$PA + PB + PC \geq \left(\frac{b}{c} + \frac{c}{b}\right) p_a + \left(\frac{c}{a} + \frac{a}{c}\right) p_b + \left(\frac{a}{b} + \frac{b}{a}\right) p_c.$$

We have the equality when P is the circumcenter O.

Second solution. Let L, M and N be the feet of the perpendicular from point P to the sides BC, CA and AB, respectively. Let H and G be the orthogonal projections of B and C, respectively, over the segment MN. Then $BC \geq HG = HN + NM + MG$.

Since $\angle BNH = \angle ANM = \angle APM$, the right triangles BNH and APM are similar, therefore $HN = \frac{PM}{PA} BN$. In an analogous way we get $MG = \frac{PN}{PA} CM$.

Applying Ptolemy's theorem to $AMPN$, we obtain $PA \cdot MN = AN \cdot PM + AM \cdot PN$, hence

$$MN = \frac{AN \cdot PM + AM \cdot PN}{PA},$$

from there we get

$$BC \geq \frac{PM}{PA} BN + \frac{AN \cdot PM + AM \cdot PN}{PA} + \frac{PN}{PA} CM.$$

Therefore,

$$BC \cdot PA \geq PM \cdot AB + PN \cdot CA.$$

Then, $PA \geq p_b \frac{c}{a} + p_c \frac{b}{a}$. Similarly for the other two inequalities.

Solution 2.73. Take a sequence of reflections of the quadrilateral $ABCD$, as shown in the figure.

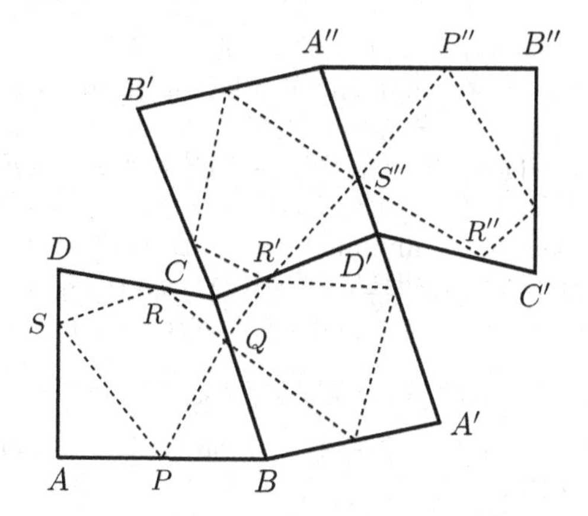

Note that the perimeter of $PQRS$ is the sum of the lengths of the piecewise line $PQR'S''P''$. Note also that $A''B''$ is parallel to AB and that the shortest distance is AA'' as can be seen if we project O on the sides of the quadrilateral.

Solution 2.74. First note that $(DEF) = (ABC) - (AFE) - (FBD) - (EDC)$.

If $x = BD$, $y = CE$, $z = AF$, $a - x = DC$, $b - y = EA$ and $c - z = FB$, we have

$$\frac{(AFE)}{(ABC)} = \frac{z(b-y)}{cb}, \; \frac{(FBD)}{(ABC)} = \frac{x(c-z)}{ac} \text{ and } \frac{(EDC)}{(ABC)} = \frac{y(a-x)}{ba}.$$

Therefore,

$$\begin{aligned}\frac{(DEF)}{(ABC)} &= 1 - \frac{z}{c}\left(1 - \frac{y}{b}\right) - \frac{x}{a}\left(1 - \frac{z}{c}\right) - \frac{y}{b}\left(1 - \frac{x}{a}\right) \\ &= \left(1 - \frac{x}{a}\right)\left(1 - \frac{y}{b}\right)\left(1 - \frac{z}{c}\right) + \frac{x}{a}\cdot\frac{y}{b}\cdot\frac{z}{c} = 2\frac{x}{a}\cdot\frac{y}{b}\cdot\frac{z}{c}.\end{aligned}$$

The last equality follows from the fact that $\frac{x}{a-x}\cdot\frac{y}{b-y}\cdot\frac{z}{c-z} = 1$ which is guaranteed because the cevians occur. Now, the last product is maximum when $\frac{x}{a} = \frac{y}{b} = \frac{z}{c}$, and since the segments concur the common value is $\frac{1}{2}$. Thus P must be the centroid.

Solution 2.75. If $x = PD$, $y = PE$ and $z = PF$, we can deduce that $2(ABC) = ax + by + cz$. Using the Cauchy-Schwarz inequality,

$$(a+b+c)^2 \leq \left(\frac{a}{x} + \frac{b}{y} + \frac{c}{z}\right)(ax + by + cz).$$

Then $\frac{a}{x} + \frac{b}{y} + \frac{c}{z} \geq \frac{(a+b+c)^2}{2(ABC)}$ and the equality holds when $x = y = z$, that is, when P is the incenter.

Solution 2.76. First, observe that $BD^2+CE^2+AF^2 = DC^2+EA^2+FB^2$, where $BD^2 - DC^2 = PB^2 - PC^2$ and similar relations have been used.

Now, $(BD + DC)^2 = a^2$, hence $BD^2 + DC^2 = a^2 - 2BD \cdot DC$. Similarly for the other two sides. Thus, $BD^2 + DC^2 + CE^2 + AE^2 + AF^2 + FB^2 = a^2 + b^2 + c^2 - 2(BD \cdot DC + CE \cdot AE + AF \cdot FB)$.

In this way, the sum is minimum when $(BD \cdot DC + CE \cdot AE + AF \cdot FB)$ is maximum. But $BD \cdot DC \leq \left(\frac{BD+DC}{2}\right)^2 = \left(\frac{a}{2}\right)^2$ and the maximum is attained when $BD = DC$. Similarly, $CE = EA$ and $AF = FB$, therefore P is the circumcenter.

Solution 2.77. Since $\sqrt[3]{(aPD)(bPE)(cPF)} \leq \frac{aPD+bPE+cPF}{3} = \frac{2(ABC)}{3}$, we can deduce that $PD \cdot PE \cdot PF \leq \frac{8}{27} \frac{(ABC)^3}{abc}$. Moreover, the equality holds if and only if $aPD = bPE = cPF$.

But $c \cdot PF = b \cdot PE \Leftrightarrow (ABP) = (CAP) \Leftrightarrow P$ is on the median AA'. Similarly, we can see that P is on the other medians, thus P is the centroid.

Solution 2.78. Using the technique for proving Leibniz's theorem, verify that $3PG^2 = PA^2 + PB^2 + PC^2 - \frac{1}{3}(a^2 + b^2 + c^2)$, where G is the centroid of ABC. Therefore, the optimal point must be $P = G$.

Solution 2.79. The quadrilateral $APMN$ is cyclic and it is inscribed in the circle of diameter AP. The chord MN always opens the angle A (or $180° - \angle A$), therefore the length of MN will depend proportionally on the radius of the circumscribed circle to $APMN$. The biggest circle will be attained when the diameter AP is the biggest possible. This happens when P is diametrally opposed to A. In this case M

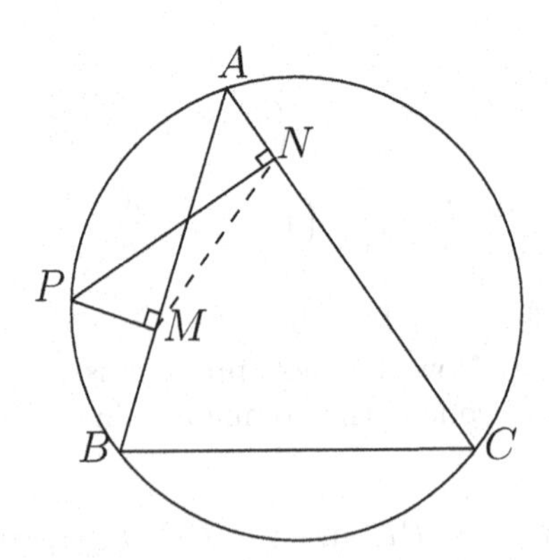

and N coincide with B and C, respectively. Therefore the maximum chord MN is BC.

Solution 2.80. The circumcircle of DEF is the nine-point circle of ABC, therefore it intersects also the midpoints of the sides of ABC and goes through L, M, N, the midpoints of AH, BH, CH, respectively.

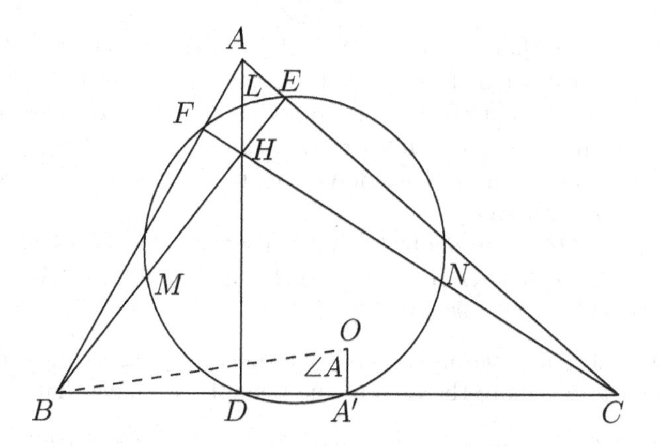

Note that $t_a^2 = AL \cdot AD$, then

$$\sum \frac{t_a^2}{h_a} = \sum \frac{AL \cdot AD}{AD} = \sum AL = \sum OA'$$
$$= \sum R\cos A \leq 3R\cos\frac{A+B+C}{3} = 3R\cos 60^\circ = \frac{3}{2}R.$$

Observe that we can prove a stronger result $\sum \frac{t_a^2}{h_a} = R + r$, using the fact that $\cos A + \cos B + \cos C = \frac{r}{R} + 1$. See Lemma 2.5.2.

Solution 2.81. (i) Notice that

$$\frac{p_a}{h_a} + \frac{p_b}{h_b} + \frac{p_c}{h_c} = \frac{ap_a}{ah_a} + \frac{bp_b}{bh_b} + \frac{cp_c}{ch_c}$$
$$= \frac{2(PBC) + 2(PCA) + 2(PAB)}{2(ABC)} = 1.$$

Now use the fact that

$$\left(\frac{p_a}{h_a} + \frac{p_b}{h_b} + \frac{p_c}{h_c}\right)\left(\frac{h_a}{p_a} + \frac{h_b}{p_b} + \frac{h_c}{p_c}\right) \geq 9.$$

(ii) Using the AM-GM inequality, we have

$$27\left(\frac{p_a}{h_a}\frac{p_b}{h_b}\frac{p_c}{h_c}\right) \leq \left(\frac{p_a}{h_a} + \frac{p_b}{h_b} + \frac{p_c}{h_c}\right)^3 = 1,$$

where the last equality follows from (i).
(iii) Let $x = (PBC)$, $y = (PCA)$ and $z = (PAB)$. Observe that $a(h_a - p_a) = ah_a - ap_a = 2(y+z) \geq 4\sqrt{yz}$. Similarly, we have that $b(h_b - p_b) \geq 4\sqrt{zx}$ y $c(h_c - p_c) \geq 4\sqrt{xy}$. Then,

$$a(h_a - p_a)b(h_b - p_b)c(h_c - p_c) \geq 64xyz = 8(ap_abp_bcp_c).$$

Therefore, $(h_a - p_a)(h_b - p_b)(h_c - p_c) \geq 8p_ap_bp_c$.

Solution 2.82. Assume that $a < b < c$, then of all the altitudes of ABC, AD is the longest. If E is the projection of I on AD, it is enough to prove that $AE \geq AO = R$. Remember that the internal bisector of $\angle A$ is also the internal bisector of $\angle EAO$. If I is projected on E' in the diameter AA', then $AE = AE'$. Now prove that $AE' \geq AO$, by proving that I is inside the acute triangle COF, where F is the intersection of AA' with BC.

To see that COF is an acute triangle, use that the angles of ABC satisfy $\angle A < \angle B < \angle C$, so that $\frac{1}{2}\angle B < 90° - \angle A$, $\frac{1}{2}\angle C < 90° - \angle A$. Use also that $\angle COF = \angle A + \angle C - \angle B < 90°$.

Solution 2.83. Let ABC be a triangle with sides of lengths a, b and c. Using Heron's formula to calculate the area of the triangle, we have that

$$(ABC) = \sqrt{s(s-a)(s-b)(s-c)}, \quad \text{where} \quad s = \frac{a+b+c}{2}. \tag{4.7}$$

If s and c are fixed, then $s - c$ is also fixed. Then the product $16(ABC)^2$ is maximum when $(s-a)(s-b)$ is maximum, that is, if $s - a = s - b$, which is equivalent to $a = b$. Therefore the triangle is isosceles.

Solution 2.84. Let ABC be a triangle with sides of length a, b and c. Since the perimeter is fixed, the semi-perimeter is also fixed. Using (4.7), we have that $16(ABC)^2$ is maximum when $(s-a)(s-b)(s-c)$ is maximum. The product of these three numbers is maximum when $(s-a) = (s-b) = (s-c)$, that is, when $a = b = c$. Therefore, the triangle is equilateral.

Solution 2.85. If a, b, c are the lengths of the sides of the triangle, observe that $a+b+c = 2R(\sin\angle A + \sin\angle B + \sin\angle C) \leq 6R\sin\left(\frac{\angle A+\angle B+\angle C}{3}\right)$, since the function $\sin x$ is concave. Moreover, equality holds when $\sin\angle A = \sin\angle B = \sin\angle C$.

Solution 2.86. The inequality $(lm + mn + nl)(l + m + n) \geq a^2 l + b^2 m + c^2 n$ is equivalent to

$$\begin{aligned}
&\frac{l^2+m^2-c^2}{lm} + \frac{m^2+n^2-b^2}{mn} + \frac{n^2+l^2-a^2}{nl} + 3 \geq 0 \\
&\Leftrightarrow \cos\angle APB + \cos\angle BPC + \cos\angle CPA + \frac{3}{2} \geq 0.
\end{aligned}$$

Now use the fact that $\cos\alpha + \cos\beta + \cos\gamma + \frac{3}{2} \geq 0$ is equivalent to $(2\cos\frac{\alpha+\beta}{2} + \cos\frac{\alpha-\beta}{2})^2 + \sin^2(\frac{\alpha-\beta}{2}) \geq 0$.

Solution 2.87. Consider the Fermat point F and let $p_1 = FA$, $p_2 = FB$ and $p_3 = FC$, then observe first that $(ABC) = \frac{1}{2}(p_1p_2 + p_2p_3 + p_3p_1)\sin 120° = \frac{\sqrt{3}}{4}(p_1p_2 + p_2p_3 + p_3p_1)$. Also,

$$\begin{aligned}
a^2 + b^2 + c^2 &= 2p_1^2 + 2p_2^2 + 2p_3^2 - 2p_1p_2\cos 120° - 2p_2p_3\cos 120° - 2p_3p_1\cos 120° \\
&= 2(p_1^2 + p_2^2 + p_3^2) + p_1p_2 + p_2p_3 + p_3p_1.
\end{aligned}$$

Now, using the fact that $x^2 + y^2 \geq 2xy$, we can deduce that $a^2 + b^2 + c^2 \geq 3(p_1p_2 + p_2p_3 + p_3p_1) = 3\left(\frac{4}{3}\sqrt{3}(ABC)\right)$. Then, $a^2 + b^2 + c^2 \geq 4\sqrt{3}(ABC)$.

Moreover, the equality $a^2 + b^2 + c^2 = 4\sqrt{3}(ABC)$ holds when $p_1^2 + p_2^2 + p_3^2 = p_1p_2 + p_2p_3 + p_3p_2$, that is, when $p_1 = p_2 = p_3$ or, equivalently, when the triangle is equilateral.

Solution 2.88. Let a, b, c be the lengths of the sides of the triangle ABC. In the same manner as we proceeded in the previous exercise, define $p_1 = FA$, $p_2 = FB$ and $p_3 = FC$. From the solution of the previous exercise we know that

$$4\sqrt{3}(ABC) = 3(p_1p_2 + p_2p_3 + p_3p_1).$$

Thus, we only need to prove that

$$3(p_1p_2 + p_2p_3 + p_3p_1) \leq (p_1 + p_2 + p_3)^2,$$

but this is equivalent to $p_1p_2 + p_2p_3 + p_3p_1 \leq p_1^2 + p_2^2 + p_3^2$, which is Exercise 1.27.

Solution 2.89. As in the Fermat problem there are two cases, when in ABC all angles are less than $120°$ or when there is an angle greater than $120°$.

In the first case the minimum of $PA + PB + PC$ is CC', where C' is the image of A when we rotate the figure in a positive direction through an angle of $60°$ having B as the center. Using the cosine law, we obtain

$$\begin{aligned}(CC')^2 &= b^2 + c^2 - 2bc\cos(A + 60°)\\ &= b^2 + c^2 - bc\cos A + bc\sqrt{3}\sin A\\ &= \frac{1}{2}(a^2 + b^2 + c^2) + 2\sqrt{3}(ABC).\end{aligned}$$

Now, use the fact that $a^2 + b^2 + c^2 \geq 4\sqrt{3}(ABC)$ to obtain $(CC')^2 \geq 4\sqrt{3}(ABC)$. Applying Theorem 2.4.3 we have that $(ABC) \geq 3\sqrt{3}r^2$, therefore $(CC')^2 \geq 36\,r^2$.

When $\angle A \geq 120°$, the point that solves Fermat-Steiner problem is the point A, then $PA + PB + PC \geq AB + AC = b + c$. It suffices to prove that $b + c \geq 6r$. Moreover, we can use the fact that $b = x + z$, $c = x + y$ and $r = \sqrt{\frac{xyz}{x+y+z}}$.

Second solution. It is clear that $PA + p_a \geq h_a$, where p_a is the distance from P to BC and h_a is the length of the altitude from A. Then $h_a + h_b + h_c \leq (PA + PB + PC) + (p_a + p_b + p_c) \leq \frac{3}{2}(PA + PB + PC)$, where the last inequality follows from Erdős-Mordell's theorem.

Now using Exercise 1.36 we have that $9 \leq (h_a + h_b + h_c)(\frac{1}{h_a} + \frac{1}{h_b} + \frac{1}{h_c}) = (h_a + h_b + h_c)(\frac{1}{r})$. Therefore, $9r \leq h_a + h_b + h_c \leq \frac{3}{2}(PA + PB + PC)$ and the result follows.

Solution 2.90. First, we note that $(A_1B_1C_1) = \frac{1}{2}A_1B_1 \cdot A_1C_1 \cdot \sin\angle B_1A_1C_1$. Since PB_1CA_1 is a cyclic quadrilateral with diameter PC, applying the sine law leads us to $A_1B_1 = PC\sin C$. Similarly, $A_1C_1 = PB\sin B$.

Call Q the intersection of BP with the circumcircle of triangle ABC, then $\angle B_1A_1C_1 = \angle QCP$. In fact, since PB_1CA_1 is a cyclic quadrilateral we have $\angle B_1CP = \angle B_1A_1P$. Similarly, $\angle C_1BP = \angle C_1A_1P$. Then $\angle B_1A_1C_1 = \angle B_1A_1P + \angle C_1A_1P = \angle B_1CP + \angle C_1BP$, but $\angle C_1BP = \angle ABQ = \angle ACQ$. Therefore, $\angle B_1A_1C_1 = \angle B_1CP + \angle ACQ = \angle QCP$.

Once again, the sine law guarantees that $\frac{\sin \angle QCP}{\sin \angle BQC} = \frac{PQ}{PC}$.

$$\begin{aligned}
(A_1B_1C_1) &= \frac{1}{2}A_1B_1 \cdot A_1C_1 \sin \angle B_1A_1C_1 \\
&= \frac{1}{2}PB \cdot PC \sin B \sin C \sin \angle QCP \\
&= \frac{1}{2}PB \cdot PC \cdot \sin B \sin C \frac{PQ}{PC} \sin \angle BQC \\
&= \frac{1}{2}PB \cdot PQ \cdot \sin A \sin B \sin C \\
&= \frac{(R^2 - OP^2)(ABC)}{4R^2}.
\end{aligned}$$

The last equality holds true because the power of the point P with respect to the circumcircle of ABC is $PB \cdot PQ = R^2 - OP^2$, and because $(ABC) = 2R^2 \sin A \sin B \sin C$. The area of $A_1B_1C_1$ is maximum when $P = O$, that is, when $A_1B_1C_1$ is the medial triangle.

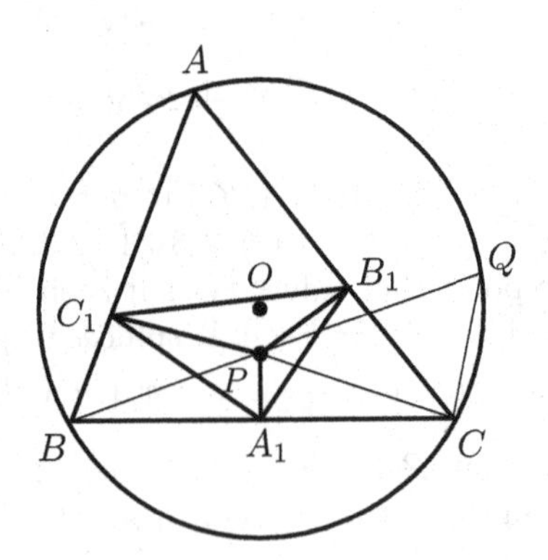

4.3 Solutions to the problems in Chapter 3

Solution 3.1. Let $a = A_1A_2$, $b = A_1A_3$ and $c = A_1A_4$. Using Ptolemy's theorem in the quadrilateral $A_1A_3A_4A_5$, we can deduce that $ab + ac = bc$ or, equivalently, $\frac{a}{b} + \frac{a}{c} = 1$.

Since the triangles $A_1A_2A_3$ and $B_1B_2B_3$ are similar, $\frac{B_1B_2}{B_1B_3} = \frac{A_1A_2}{A_1A_3} = \frac{a}{b}$ and from there we obtain $B_1B_2 = \frac{a^2}{b}$. Similarly $C_1C_2 = \frac{a^2}{c}$. Therefore $\frac{S_B+S_C}{S_A} = \frac{a^2}{b^2} + \frac{a^2}{c^2} = \frac{a^2c^2+a^2b^2}{b^2c^2} = \frac{b^2+c^2}{(b+c)^2} > \frac{(b+c)^2}{2(b+c)^2} = \frac{1}{2}$. The third equality follows from $ab + ac = bc$ and the inequality follows from inequality (1.11). The inequality is strict since $b \neq c$.

Note that $\frac{a^2}{b^2} + \frac{a^2}{c^2} = \left(\frac{a}{b} + \frac{a}{c}\right)^2 - 2\frac{a^2}{bc} = 1 - 2\frac{a^2}{bc}$.

The sine law applied to the triangle $A_1A_3A_4$ leads us to

$$\begin{aligned}
\frac{a^2}{bc} &= \frac{\sin^2 \frac{\pi}{7}}{\sin \frac{2\pi}{7} \sin \frac{4\pi}{7}} = \frac{\sin^2 \frac{\pi}{7}}{2 \sin \frac{2\pi}{7} \sin \frac{2\pi}{7} \cos \frac{2\pi}{7}} \\
&= \frac{\sin^2 \frac{\pi}{7}}{2(1 - \cos^2 \frac{2\pi}{7}) \cos \frac{2\pi}{7}} = \frac{\sin^2 \frac{\pi}{7}}{2 \cos \frac{2\pi}{7} (1 + \cos \frac{2\pi}{7})(1 - \cos \frac{2\pi}{7})} \\
&= \frac{\sin^2 \frac{\pi}{7}}{4 \cos \frac{2\pi}{7} (1 + \cos \frac{2\pi}{7}) \sin^2 \frac{\pi}{7}} = \frac{1}{4 \cos \frac{2\pi}{7} (1 + \cos \frac{2\pi}{7})} \\
&> \frac{1}{4 \cos \frac{\pi}{4} (1 + \cos \frac{\pi}{4})} = \frac{1}{4\frac{\sqrt{2}}{2}(1 + \frac{\sqrt{2}}{2})} = \frac{\sqrt{2} - 1}{2}.
\end{aligned}$$

Thus $\frac{a^2}{b^2} + \frac{a^2}{c^2} = 1 - 2\frac{a^2}{bc} < 1 - (\sqrt{2} - 1) = 2 - \sqrt{2}$.

Solution 3.2. Cut the tetrahedron along the edges AD, BD, CD and place it on the plane of the triangle ABC. The faces ABD, BCD and CAD will have as their image the triangles ABD_1, BCD_2 and CAD_3. Observe that D_3, A and D_1 are collinear, as are D_1, B and D_2. Moreover, A is the midpoint of D_1D_3 (since both D_1A and D_3A are equal in length to DA), and similarly B is the midpoint of D_1D_2. Then $AB = \frac{1}{2}D_2D_3$ and by the triangle inequality, $D_2D_3 \leq CD_3 + CD_2 = 2CD$. Hence $AB \leq CD$, as desired.

Solution 3.3. Letting S be the area of the triangle, we have the formulae $\sin \alpha = \frac{2S}{bc}$, $\sin \beta = \frac{2S}{ca}$, $\sin \gamma = \frac{2S}{ab}$ and $r = \frac{S}{s} = \frac{2S}{a+b+c}$. Using these formulae we find that the inequality to be proved is equivalent to

$$\left(\frac{a}{bc} + \frac{b}{ca} + \frac{c}{ab}\right)(a + b + c) \geq 9,$$

which can be proved by applying the AM-GM inequality to each factor on the left side.

Solution 3.4. Suppose that the circles have radii 1. Let P be the common point of the circles and let A, B, C be the second intersection points of each pair of circles. We have to minimize the common area between any pair of circles, which will be minimum if the point P is in the interior of the triangle ABC (otherwise, rotate one circle by $180°$ around P, and this will reduce the common area).

The area of the common parts is equal to $\pi - (\sin \alpha + \sin \beta + \sin \gamma)$, where α, β, γ are the central angles of the common arcs of the circles. It is clear that

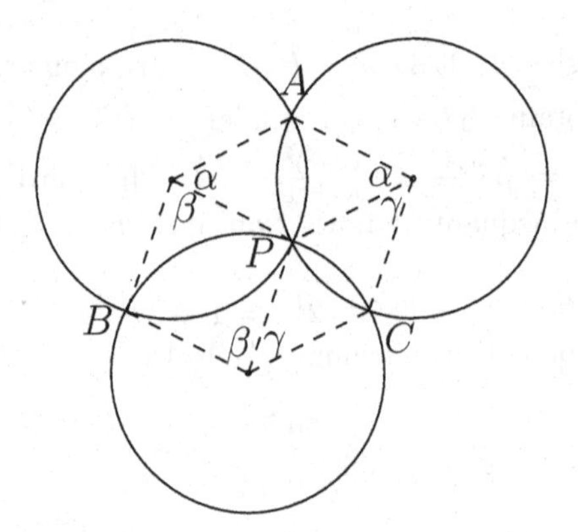

$\alpha+\beta+\gamma = 180°$. Since the function $\sin x$ is concave, the minimum is reached when $\alpha = \beta = \gamma = \frac{\pi}{3}$, which implies that the centers of the circles form an equilateral triangle.

Solution 3.5. Let I be the incenter of ABC, and draw the line through I perpendicular to IC. Let D', E' be the intersections of this line with BC, CA, respectively. First prove that $(CDE) \geq (CD'E')$ by showing that the area of $D'DI$ is greater than or equal to the area of $EE'I$; to see this, observe that one of the triangles $DD'I$, $EE'I$ lies in the opposite side to C with respect to the line $D'E'$, if for instance, it is $DD'I$, then this triangle will have a greater area than the area of $EE'I$, then the claim.

Now, prove that the area $(CD'E')$ is $\frac{2r^2}{\sin C}$; to see this, note that $CI = \frac{r}{\sin \frac{C}{2}}$ and that $D'I = \frac{r}{\cos \frac{C}{2}}$, then

$$(CD'E') = \frac{1}{2}D'E' \cdot CI = D'I \cdot CI = \frac{2r^2}{2\sin \frac{C}{2} \cos \frac{C}{2}} = \frac{2r^2}{\sin C} \geq 2r^2.$$

Solution 3.6. The key is to note that $2AX \geq \sqrt{3}(AB+BX)$, which can be deduced by applying Ptolemy's inequality (Exercise 2.11) to the cyclic quadrilateral $ABXO$ that is formed when we glue the triangle ABX to the equilateral triangle AXO of side AX, and then observing that the diameter of the circumcircle of the equilateral triangle is $\frac{2}{\sqrt{3}}AX$, that is, $AX(AB + BX) = AX \cdot BO \leq AX \cdot \frac{2}{\sqrt{3}}AX$. Hence

$$\begin{aligned} 2AD = 2(AX + XD) &\geq \sqrt{3}(AB + BX) + 2XD \\ &\geq \sqrt{3}(AB + BC + CX) + \sqrt{3}XD \\ &\geq \sqrt{3}(AB + BC + CD). \end{aligned}$$

Solution 3.7. Take the triangle $A'B'C'$ of maximum area between all triangles that can be formed with three points of the given set of points; then its area satisfies $(A'B'C') \leq 1$. Construct another triangle ABC that has $A'B'C'$ as its medial triangle; this has an area $(ABC) = 4(A'B'C') \leq 4$. In ABC we can find all the points. Indeed, if some point Q is outside of ABC, it will be in one of the half-planes determined by the sides of the triangle and opposite to the half-plane where the third vertex lies. For instance, if Q is in the half-plane determined by BC, opposite to where A lies, the triangle $QB'C'$ has greater area than $A'B'C'$, a contradiction.

Solution 3.8. Let $M = 1+\frac{1}{2}+\cdots+\frac{1}{n}$. Let us prove that M is the desired minimum value, which is achieved by setting $x_1 = x_2 = \cdots = x_n = 1$. Using the AM-GM inequality, we get $x_k^k + (k-1) \geq kx_k$ for all k. Therefore

$$x_1+\frac{x_2^2}{2}+\frac{x_3^3}{3}+\cdots+\frac{x_n^n}{n} \geq x_1+x_2-\frac{1}{2}+\cdots+x_n-\frac{n-1}{n} = x_1+x_2+\cdots+x_n-n+M.$$

On the other hand, the arithmetic-harmonic inequality leads us to

$$\frac{x_1+x_2+\cdots+x_n}{n} \geq \frac{n}{\frac{1}{x_1}+\frac{1}{x_1}+\cdots+\frac{1}{x_n}} = 1.$$

We conclude that the given expression is at least $n-n+M = M$. Since M is achieved, it is the desired minimum.

Second solution. Apply the weighted AM-GM inequality to the numbers $\{x_j^j\}$ with weights $\left\{t_j = \frac{\frac{1}{j}}{\sum\frac{1}{j}}\right\}$, to get

$$\sum\frac{x_j^j}{j} \geq \left(\sum\frac{1}{j}\right)(x_1x_2\cdots x_n)^{\frac{1}{\sum\frac{1}{j}}} \geq \sum\frac{1}{j}.$$

The last inequality follows from $n\sqrt[n]{\frac{1}{x_1}\cdots\frac{1}{x_n}} \leq \sum\frac{1}{x_j} = n$.

Solution 3.9. Note that AFE and BDC are equilateral triangles. Let C' and F' be points outside the hexagon and such that ABC' and DEF' are also equilateral triangles. Since BE is the perpendicular bisector of AD, it follows that C' and F' are the reflections of C and F on the line BE. Now use the fact that $AC'BG$ and $EF'DH$ are cyclic in order to conclude that $AG+GB = GC'$ and $DH+HE = HF'$.

Solution 3.10. Leibniz's theorem implies $OG^2 = R^2 - \frac{1}{9}(a^2+b^2+c^2)$. Since $rs = \frac{abc}{4R}$, we can deduce that $2rR = \frac{abc}{a+b+c}$. Then we have to prove that $abc \leq \frac{(a+b+c)}{3}\frac{(a^2+b^2+c^2)}{3}$, for which we can use the AM-GM inequality.

Solution 3.11. The left-hand side of the inequality follows from

$$\sqrt{1+x_0+x_1+\cdots+x_{i-1}}\sqrt{x_i+\cdots+x_n} \leq \frac{1}{2}(1+x_0+\cdots+x_n) = 1.$$

For the right-hand side consider $\theta_i = \arcsin(x_0 + \cdots + x_i)$ for $i = 0, \ldots, n$. Note that

$$\sqrt{1 + x_0 + \cdots + x_{i-1}}\sqrt{x_i + \cdots + x_n} = \sqrt{1 + \sin\theta_{i-1}}\sqrt{1 - \sin\theta_{i-1}}$$
$$= \cos\theta_{i-1}.$$

It is left to prove that $\sum \frac{\sin\theta_i - \sin\theta_{i-1}}{\cos\theta_{i-1}} < \frac{\pi}{2}$. But

$$\sin\theta_i - \sin\theta_{i-1} = 2\cos\frac{\theta_i + \theta_{i-1}}{2}\sin\frac{\theta_i - \theta_{i-1}}{2} < (\cos\theta_{i-1})(\theta_i - \theta_{i-1}).$$

To show the inequality, use the facts that $\cos\theta$ is a decreasing function and that $\sin\theta \leq \theta$ for $0 \leq \theta \leq \frac{\pi}{2}$. Then

$$\sum \frac{\sin\theta_i - \sin\theta_{i-1}}{\cos\theta_{i-1}} < \sum \theta_i - \theta_{i-1} = \theta_n - \theta_0 = \frac{\pi}{2}.$$

Solution 3.12. If $\sum_{i=1}^n x_i = 1$, then $1 = \left(\sum_{i=1}^n x_i\right)^2 = \sum_{i=1}^n x_i^2 + 2\sum_{i<j} x_i x_j$. Therefore the inequality that we need to prove is equivalent to

$$\frac{1}{n-1} \leq \sum_{i=1}^n \frac{x_i^2}{1 - a_i}.$$

Use the Cauchy-Schwarz inequality to prove that

$$\left(\sum_{i=1}^n x_i\right)^2 \leq \sum_{i=1}^n \frac{x_i^2}{1 - a_i} \sum_{i=1}^n (1 - a_i).$$

Solution 3.13. First prove that $\sum_{i=1}^n x_{n+1}(x_{n+1} - x_i) = (n-1)x_{n+1}^2$. The inequality that we need to prove is reduced to

$$\sum_{i=1}^n \sqrt{x_i(x_{n+1} - x_i)} \leq \sqrt{n-1}x_{n+1}.$$

Now use the Cauchy-Schwarz inequality with the following two n-sets of real numbers: $(\sqrt{x_1}, \ldots, \sqrt{x_n})$ and $(\sqrt{x_{n+1} - x_1}, \ldots, \sqrt{x_{n+1} - x_n})$.

Solution 3.14. First, recall that N is also the midpoint of the segment that joins the midpoints X and Y of the diagonals AC and BD. The circle of diameter OM goes through X and Y since OX and OY are perpendiculars to the corresponding diagonals, and ON is a median of the triangle OXY.

Solution 3.15. The inequality on the right-hand side follows from $wx + xy + yz + zw = (w + y)(x + z) = -(w + y)^2 \leq 0$.

For the inequality on the left-hand side, note that

$$\begin{aligned}|wx + xy + yz + zw| &= |(w + y)(x + z)| \\ &\leq \frac{1}{2}\left[(w + y)^2 + (x + z)^2\right] \\ &\leq w^2 + x^2 + y^2 + z^2 = 1.\end{aligned}$$

We can again use the Cauchy-Schwarz inequality to obtain

$$|wx + xy + yz + zw|^2 \leq (w^2 + x^2 + y^2 + z^2)(x^2 + y^2 + z^2 + w^2) = 1.$$

Solution 3.16. For the inequality on the left-hand side, rearrange as follows:

$$\frac{a_n + a_2}{a_1} + \frac{a_1 + a_3}{a_2} + \cdots + \frac{a_{n-1} + a_1}{a_n} = \frac{a_1}{a_2} + \frac{a_2}{a_1} + \frac{a_3}{a_2} + \frac{a_2}{a_3} + \cdots + \frac{a_1}{a_n} + \frac{a_n}{a_1},$$

now, use the fact that $\left(\frac{x}{y} + \frac{y}{x}\right) \geq 2$.

Set $S_n = \frac{a_n+a_2}{a_1} + \frac{a_1+a_3}{a_2} + \frac{a_2+a_4}{a_3} + \cdots + \frac{a_{n-1}+a_1}{a_n}$. Using induction, prove that $S_n \leq 3n$.

First, for $n = 3$, we need to see that $\frac{b+c}{a} + \frac{c+a}{b} + \frac{a+b}{c} \leq 9$. If $a = b = c$, then $\frac{b+c}{a} + \frac{c+a}{b} + \frac{a+b}{c} = 6$ and the inequality is true. Suppose that $a \leq b \leq c$ and that not all numbers are equal, then we have three cases: $a = b < c$, $a < b = c$, $a < b < c$. In all of them, we have $a \leq b$ and $a < c$. Hence $2c = c + c > a + b$ and $\frac{a+b}{c} < 2$, and since $\frac{a+b}{c}$ is a positive integer we have $c = a + b$.

Thus, $\frac{b+c}{a} + \frac{c+a}{b} + \frac{a+b}{c} = \frac{a+2b}{a} + \frac{2a+b}{b} + 1 = 3 + 2\frac{b}{a} + 2\frac{a}{b}$. Since $2\frac{b}{a}$ and $2\frac{a}{b}$ are positive integers, and since $\left(2\frac{b}{a}\right)\left(2\frac{a}{b}\right) = 4$, we have that either both numbers are equal to 2 or one number is 1 and the other is 4. This means the sum is at most 8, which is less than 9, then the result.

We continue with the induction. Suppose that $S_{n-1} \leq 3(n-1)$. Consider $\{a_1, \ldots, a_n\}$, if all are equal, then $S_n = 2n$ and the inequality is true. Suppose instead that there are at least two differents a_i's. Take the maximum of the a_i's; its neighbors (a_{i-1}, a_{i+1}) can be equal to this maximum value, but since there are two different numbers between the a_i's for some maximum a_i, we have that one of its neighbors is less than a_i. We can then assume, without loss of generality, that a_n is maximum and that one of its neighbors, a_{n-1} or a_1, is less than a_n. Then, since $2a_n > a_{n-1} + a_1$, we have that $\frac{a_{n-1}+a_1}{a_n} < 2$ and then $\frac{a_{n-1}+a_1}{a_n} = 1$, for which $a_n = a_{n-1} + a_1$. When we substitute this value of a_n in S_n, we get

$$\begin{aligned}S_n &= \frac{a_{n-1} + a_1 + a_2}{a_1} + \frac{a_2 + a_3}{a_2} + \cdots + \frac{a_{n-2} + a_{n-1} + a_1}{a_{n-1}} + \frac{a_{n-1} + a_1}{a_{n-1} + a_1} \\ &= 1 + \frac{a_{n-1} + a_2}{a_1} + \frac{a_2 + a_3}{a_2} + \cdots + \frac{a_{n-2} + a_1}{a_{n-1}} + 1 + 1.\end{aligned}$$

Since $S_{n-1} \leq 3(n-1)$, this implies that $S_n \leq 3n$.

Solution 3.17. Since the quadrilateral $OBDC$ is cyclic, use Ptolemy's theorem to prove that $OD = R\left(\frac{BD}{BC} + \frac{DC}{BC}\right)$, where R is the circumradius of ABC. On the other hand, since the triangles BCE and DCA are similar, as well as the triangles ABD and FBC, it happens that $R\left(\frac{BD}{BC} + \frac{DC}{BC}\right) = R\left(\frac{AD}{FC} + \frac{AD}{EB}\right)$. We can find similar equalities for OE and OF, $OE = R\left(\frac{BE}{AD} + \frac{BE}{CF}\right)$ and $OF = R\left(\frac{CF}{BE} + \frac{CF}{AD}\right)$. Multiplying these equalities and applying the AM-GM inequality, the result is attained.

Another way to prove this is using inversion. Let D', E' and F' be the intersection points of AO, BO and CO with the sides BC, CA and AB, respectively. Invert the sides BC, CA and AB with respect to (O, R), obtaining the circumcircles of the triangles OBC, OCA and OAB, respectively. Then, $OD \cdot OD' = OE \cdot OE' = OF \cdot OF' = R^2$. If $x = (ABO)$, $y = (BCO)$ and $z = (CAO)$, we can deduce that

$$\frac{AO}{OD'} = \frac{z+x}{y}, \quad \frac{BO}{OE'} = \frac{x+y}{z} \quad \text{and} \quad \frac{CO}{OF'} = \frac{y+z}{x}.$$

This implies, using the AM-GM inequality, that $\frac{R^3}{OD' \cdot OE' \cdot OF'} \geq 8$; therefore, $OD \cdot OE \cdot OF \geq 8R^3$.

Solution 3.18. First, observe that $AY \leq 2R$ and that $h_a \leq AX$, where h_a is the length of the altitude on BC. Then we can deduce that

$$\begin{aligned}
\sum \frac{l_a}{\sin^2 A} &= \sum \frac{AX}{AY \sin^2 A} \\
&\geq \sum \frac{h_a}{2R \sin^2 A} \\
&= \sum \frac{h_a}{a \sin A} \quad \left(\text{since } \frac{\sin A}{a} = \frac{1}{2R}\right) \\
&\geq 3\sqrt[3]{\frac{h_a}{a \sin A} \frac{h_b}{b \sin B} \frac{h_c}{c \sin C}} \\
&= 3
\end{aligned}$$

since $h_a = b \sin C$, $h_b = c \sin A$, $h_c = a \sin B$.

Solution 3.19. Without loss of generality, $x_1 \leq x_2 \leq \cdots \leq x_n$. Since $1 < 2 < \cdots < n$, we have, using the rearrangement inequality (1.2), that

$$A = x_1 + 2x_2 + \cdots + nx_n \geq nx_1 + (n-1)x_2 + \cdots + x_n = B.$$

Then, $|A + B| = |(n+1)(x_1 + \cdots + x_n)| = n + 1$, hence $A + B = \pm(n+1)$. Now, if $A + B = (n+1)$ it follows that $B \leq \frac{n+1}{2} \leq A$, and if $A + B = -(n+1)$, it is the case that $B \leq -\frac{n+1}{2} \leq A$

If we now assume that $\frac{n+1}{2}$ or $-\frac{n+1}{2}$ is between B and A, otherwise A or B would be in the interval $\left[-\frac{n+1}{2}, \frac{n+1}{2}\right]$, then either $|A|$ or $|B|$ is less than or equal to $\frac{n+1}{2}$ and we can solve the problem.

Suppose therefore that $B \le -\frac{n+1}{2} < \frac{n+1}{2} \le A$.

Let $y_1, \ldots, y_n$ be a permutation of $x_1, \ldots, x_n$ such that $1y_1+2y_2+\cdots+ny_n = C$ takes the maximum value with $C \le -\frac{n+1}{2}$. Take i such that $y_1 \le y_2 \le \cdots \le y_i$ and $y_i > y_{i+1}$ and consider

$$D = y_1 + 2y_2 + \cdots + iy_{i+1} + (i+1)y_i + (i+2)y_{i+2} + \cdots + ny_n$$
$$D - C = iy_{i+1} + (i+1)y_i - (iy_i + (i+1)y_{i+1}) = y_i - y_{i+1} > 0.$$

Since $|y_i|$, $|y_{i+1}| \le \frac{n+1}{2}$, we can deduce that $D - C = y_i - y_{i+1} \le n+1$; hence $D \le C + n + 1$ and therefore $C < D \le C + n + 1 \le \frac{n+1}{2}$.

On the other hand, $D \ge -\frac{n+1}{2}$, since C is the maximum sum which is less than $-\frac{n+1}{2}$. Thus $-\frac{n+1}{2} \le D \le \frac{n+1}{2}$ and then $|D| \le \frac{n+1}{2}$.

Solution 3.20. Among the numbers x, y, z two have the same sign (say x and y), since $c = z\left(\frac{x}{y} + \frac{y}{x}\right)$ is positive, we can deduce that z is positive.

Note that $a+b-c = \frac{2xy}{z}$, $b+c-a = \frac{2yz}{x}$, $c+a-b = \frac{2zx}{y}$ are positive.

Conversely, if $u = a+b-c$, $v = b+c-a$ and $w = c+a-b$ are positive, taking $u = \frac{2xy}{z}$, $v = \frac{2yz}{x}$, $w = \frac{2zx}{y}$, we can obtain $a = \frac{u+w}{2} = x\left(\frac{y}{z} + \frac{z}{y}\right)$, and so on.

Solution 3.21. First, prove that a centrally symmetric hexagon $ABCDEF$ has opposite parallel sides. Thus, $(ACE) = (BDF) = \frac{(ABCDEF)}{2}$. Now, if we reflect the triangle PQR with respect to the symmetry center of the hexagon, we get the points P', Q', R' which form the centrally symmetric hexagon $PR'QP'RQ'$, inscribed in $ABCDEF$ with area $2(PQR)$.

Solution 3.22. Let $X = \sum_{i=1}^4 x_i^3$, $X_i = X - x_i^3$; it is then evident that $X = \frac{1}{3}\sum_{i=1}^4 X_i$. Using the AM-GM inequality leads to $\frac{1}{3}X_1 \ge \sqrt[3]{x_2^3x_3^3x_4^3} = \frac{1}{x_1}$; similar inequalities hold for the other indexes and this implies that $X \ge \sum_{i=1}^4 \frac{1}{x_i}$.

Using Tchebyshev's inequality we obtain

$$\frac{x_1^3 + x_2^3 + x_3^3 + x_4^3}{4} \ge \frac{x_1^2 + x_2^2 + x_3^2 + x_4^2}{4} \cdot \frac{x_1 + x_2 + x_3 + x_4}{4}.$$

Thanks to the AM-GM inequality we get $\frac{x_1^2+x_2^2+x_3^2+x_4^2}{4} \ge \sqrt[4]{(x_1x_2x_3x_4)^2} = 1$, and therefore $X \ge \sum_{i=1} x_i$.

Solution 3.23. Use the Cauchy-Schwarz inequality with $u = \left(\frac{\sqrt{x-1}}{\sqrt{x}}, \frac{\sqrt{y-1}}{\sqrt{y}}, \frac{\sqrt{z-1}}{\sqrt{z}}\right)$ and $v = \left(\sqrt{x}, \sqrt{y}, \sqrt{z}\right)$.

Solution 3.24. If $\alpha = \angle ACM$ and $\beta = \angle BDM$, then $\frac{MA \cdot MB}{MC \cdot MD} = \tan\alpha\tan\beta$ and $\alpha + \beta = \frac{\pi}{4}$. Now use the fact that $\tan\alpha\tan\beta\tan\gamma \le \tan^3\left(\frac{\alpha+\beta+\gamma}{3}\right)$, where $\gamma = \frac{\pi}{4}$.

Another method uses the fact that the inequality is equivalent to $(MCD) \ge 3\sqrt{3}(MAB)$ which is equivalent to $\frac{h+l}{h} \ge 3\sqrt{3}$, where l is the length of the side of the square and h is the length of the altitude from M to AB. Find the maximum h.

Solution 3.25. First note that $\frac{PL}{AL} + \frac{PM}{BM} + \frac{PN}{CN} = 1$. Now, use the fact that AL, BM and CN are less than a.

Solution 3.26. Since $\frac{PB}{PA} \cdot \frac{QC}{QA} \leq \frac{1}{4}\left(\frac{PB}{PA} + \frac{QC}{QA}\right)^2$, it is sufficient to see that $\frac{PB}{PA} + \frac{QC}{QA} = 1$.

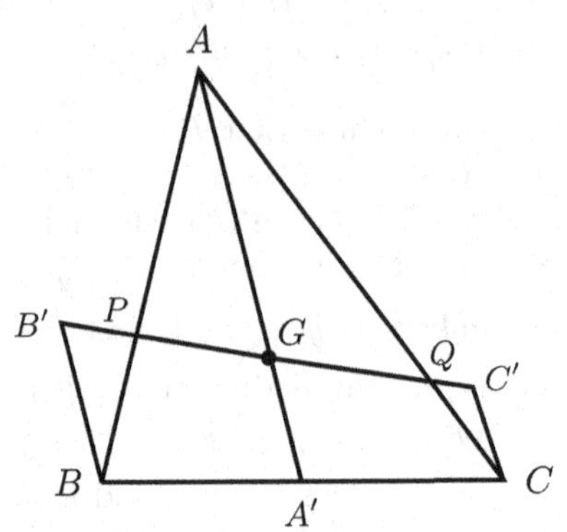

Draw BB', CC' parallel to the median AA' in such a way that B' and C' are on PQ. The triangles APG and BPB' are similar, as well as AQG and CQC', thus $\frac{PB}{PA} = \frac{BB'}{AG}$ and $\frac{QC}{QA} = \frac{CC'}{AG}$. Use this together with the fact that $AG = 2GA' = BB' + CC'$.

Solution 3.27. Let Γ be the circumcircle of ABC, and let R be its radius. Consider the inversion in Γ. For any point P other than O, let P' be its inverse. The inverse of the circumcircle of OBC is the line BC, then A_1', the inverse of A_1, is the intersection point between the ray OA_1 and BC. Since[22]

$$P'Q' = \frac{R^2 \cdot PQ}{OP \cdot OQ}$$

for two points P, Q (distinct from O) with inverses P', Q', we have

$$\frac{AA_1}{OA_1} = \frac{R^2 \cdot A'A_1'}{OA' \cdot OA_1' \cdot OA_1} = \frac{AA_1'}{OA} = \frac{x+y+z}{y+z},$$

where x, y, z denote the areas of the triangles OBC, OCA, OAB, respectively. Similarly, we have that

$$\frac{BB_1}{OB_1} = \frac{x+y+z}{z+x} \quad \text{and} \quad \frac{CC_1}{OC_1} = \frac{x+y+z}{x+y}.$$

Thus

$$\frac{AA_1}{OA_1} + \frac{BB_1}{OB_1} + \frac{CC_1}{OC_1} = (x+y+z)\left(\frac{1}{y+z} + \frac{1}{z+x} + \frac{1}{x+y}\right) \geq \frac{9}{2}.$$

For the last inequality, see Exercise 1.44.

[22] See [5, page 132] or [9, page 112].

Solution 3.28. The area of the triangle GBC is $(GBC) = \frac{(ABC)}{3} = \frac{a \cdot GL}{2}$. Therefore $GL = \frac{2(ABC)}{3a}$. Similarly, $GN = \frac{2(ABC)}{3c}$.

In consequence,

$$\begin{aligned}(GNL) &= \frac{GL \cdot GN \sin B}{2} = \frac{4(ABC)^2 \sin B}{18ac} \\ &= \frac{4(ABC)^2 b^2}{(18abc)(2R)} = \frac{(ABC)^2\, b^2}{(9R\,\frac{abc}{4R})(4R)} \\ &= \frac{(ABC)\, b^2}{9 \cdot 4R^2}.\end{aligned}$$

Similarly, $(GLM) = \frac{(ABC)\, c^2}{9 \cdot 4R^2}$ and $(GMN) = \frac{(ABC)\, a^2}{9 \cdot 4R^2}$. Therefore,

$$\frac{(LMN)}{(ABC)} = \frac{1}{9}\left(\frac{a^2+b^2+c^2}{4R^2}\right) = \frac{R^2 - OG^2}{4R^2}.$$

The inequality in the right follows easily.

For the other inequality, note that $OG = \frac{1}{3}OH$. Since the triangle is acute, H is inside the triangle and $HO \leq R$. Therefore,

$$\frac{(LMN)}{(ABC)} = \frac{R^2 - \frac{1}{9}OH^2}{4R^2} \geq \frac{R^2 - \frac{1}{9}R^2}{4R^2} = \frac{2}{9} > \frac{4}{27}.$$

Solution 3.29. The function $f(x) = \frac{1}{1+x}$ is convex for $x > 0$. Thus,

$$\begin{aligned}\frac{f(ab)+f(bc)+f(ca)}{3} &\geq f\left(\frac{ab+bc+ca}{3}\right) = \frac{3}{3+ab+bc+ca} \\ &\geq \frac{3}{3+a^2+b^2+c^2} = \frac{1}{2},\end{aligned}$$

the last inequality follows from $ab+bc+ca \leq a^2+b^2+c^2$.

We can also begin with

$$\frac{1}{1+ab} + \frac{1}{1+bc} + \frac{1}{1+ca} \geq \frac{9}{3+ab+bc+ca} \geq \frac{9}{3+a^2+b^2+c^2} = \frac{3}{2}.$$

The first inequality follows from inequality (1.11) and the second from Exercise 1.27.

Solution 3.30. Set $x = b+2c$, $y = c+2a$, $z = a+2b$. The desired inequality becomes

$$\left(\frac{x}{y}+\frac{y}{x}\right) + \left(\frac{y}{z}+\frac{z}{y}\right) + \left(\frac{z}{x}+\frac{x}{z}\right) + 3\left(\frac{y}{x}+\frac{z}{y}+\frac{x}{z}\right) \geq 15,$$

which can be proved using the AM-GM inequality. Another way of doing it is the following:

$$\frac{a}{b+2c}+\frac{b}{c+2a}+\frac{c}{a+2b}=\frac{a^2}{ab+2ca}+\frac{b^2}{bc+2ab}+\frac{c^2}{ca+2bc}\geq\frac{(a+b+c)^2}{3(ab+bc+ca)}.$$

The inequality follows from inequality (1.11). It remains to prove the inequality $(a+b+c)^2 \geq 3(ab+bc+ca)$, which is a consequence of the Cauchy-Schwarz inequality.

Solution 3.31. Use the inequality (1.11) or use the Cauchy-Schwarz inequality with

$$\left(\frac{a}{\sqrt{a+b}},\frac{b}{\sqrt{b+c}},\frac{c}{\sqrt{c+d}},\frac{d}{\sqrt{d+a}}\right) \quad \text{and} \quad (\sqrt{a+b},\sqrt{b+c},\sqrt{c+d},\sqrt{d+a}).$$

Solution 3.32. Let $x=b+c-a$, $y=c+a-b$ and $z=a+b-c$. The similarity between the triangles ADE and ABC gives us

$$\frac{DE}{a}=\frac{\text{perimeter of } ADE}{\text{perimeter of } ABC}=\frac{2x}{a+b+c}.$$

Thus, $DE=\frac{x(y+z)}{x+y+z}$; that is, the inequality is equivalent to $\frac{x(y+z)}{x+y+z}\leq\frac{x+y+z}{4}$. Now use the AM-GM inequality.

Solution 3.33. Take F on AD with $AF=BC$ and define E' as the intersection of BF and AC. Using the sine law in the triangles $AE'F$, BCE' and BDF, we obtain

$$\frac{AE'}{E'C}=\frac{AF\sin F}{\sin E'}\cdot\frac{\sin E'}{BC\sin B}=\frac{\sin F}{\sin B}=\frac{BD}{FD}=\frac{AE}{EC},$$

therefore $E'=E$.

Subsequently, consider G on BD with $BG=AD$ and H the intersection point of GE with the parallel to BC passing through A. Use the fact that the triangles ECG and EAH are similar and also use Menelaus's theorem for the triangle CAD with transversal EFB to conclude that $AH=DB$. Hence, $BDAH$ is a parallelogram, $BH=AD$ and BHG is isosceles with $BH=BG=AD>BE$.

Solution 3.34. Note that $ab+bc+ca\leq 3abc$ if and only if $\frac{1}{a}+\frac{1}{b}+\frac{1}{c}\leq 3$. Since

$$(a+b+c)\left(\frac{1}{a}+\frac{1}{b}+\frac{1}{c}\right)\geq 9,$$

we should have that $(a+b+c)\geq 3$. Then

$$\begin{aligned}3(a+b+c)&\leq(a+b+c)^2\\&=\left(a^{3/2}a^{-1/2}+b^{3/2}b^{-1/2}+c^{3/2}c^{-1/2}\right)^2\\&\leq\left(a^3+b^3+c^3\right)\left(\frac{1}{a}+\frac{1}{b}+\frac{1}{c}\right)\\&\leq 3\left(a^3+b^3+c^3\right).\end{aligned}$$

Solution 3.35. Take $y_i = \frac{x_i}{n-1}$ for all $i = 1, 2, \ldots, n$ and suppose that the inequality is false, that is,

$$\frac{1}{1+y_1} + \frac{1}{1+y_2} + \cdots + \frac{1}{1+y_n} > n-1.$$

Then

$$\begin{aligned}\frac{1}{1+y_i} &> \sum_{j\neq i}\left(1 - \frac{1}{1+y_j}\right) = \sum_{j\neq i}\frac{y_j}{1+y_j}\\ &\geq (n-1)\sqrt[n-1]{\frac{y_1\cdots\hat{y_i}\cdots y_n}{(1+y_1)\cdots\widehat{(1+y_i)}\cdots(1+y_n)}},\end{aligned}$$

where $y_1 \cdots \hat{y_i} \cdots y_n$ is the product of the y's except y_i. Then

$$\prod_{i=1}^{n}\frac{1}{1+y_i} > (n-1)^n\frac{y_1\cdots y_n}{(1+y_1)\cdots(1+y_n)},$$

and this implies $1 > x_1 \cdots x_n$, a contradiction.

Solution 3.36. Use the Cauchy-Schwarz inequality with $\left(\sqrt{\frac{x_1}{y_1}}, \ldots, \sqrt{\frac{x_n}{y_n}}\right)$ and $\left(\sqrt{x_1y_1}, \ldots, \sqrt{x_ny_n}\right)$ to get

$$\begin{aligned}(x_1 + \cdots + x_n)^2 &= \left(\sqrt{\frac{x_1}{y_1}}\sqrt{x_1y_1} + \cdots + \sqrt{\frac{x_n}{y_n}}\sqrt{x_ny_n}\right)^2\\ &\leq \left(\frac{x_1}{y_1} + \cdots + \frac{x_n}{y_n}\right)(x_1y_1 + \cdots + x_ny_n).\end{aligned}$$

Now use the hypothesis $\sum x_iy_i \leq \sum x_i$.

Solution 3.37. Since $abc = 1$, we have

$$(a-1)(b-1)(c-1) = a+b+c-\left(\frac{1}{a}+\frac{1}{b}+\frac{1}{c}\right)$$

and similarly

$$(a^n-1)(b^n-1)(c^n-1) = a^n+b^n+c^n-\left(\frac{1}{a^n}+\frac{1}{b^n}+\frac{1}{c^n}\right).$$

The proof follows from the fact that the left sides of the equalities have the same sign.

Solution 3.38. We prove the claim using induction on n. The case $n = 1$ is clear. Now assuming the claim is true for n, we can prove it is true for $n+1$.
Since $n < \sqrt{n^2+i} < n+1$, for $i = 1, 2, \ldots, 2n$, we have

$$\left\{\sqrt{n^2+i}\right\} = \sqrt{n^2+i} - n < \sqrt{n^2+i+\left(\frac{i}{2n}\right)^2} - n = \frac{i}{2n}.$$

Thus

$$\sum_{j=1}^{(n+1)^2}\left\{\sqrt{j}\right\}=\sum_{j=1}^{n^2}\left\{\sqrt{j}\right\}+\sum_{j=n^2+1}^{(n+1)^2}\left\{\sqrt{j}\right\}\le\frac{n^2-1}{2}+\frac{1}{2n}\sum_{i=1}^{2n}i$$
$$=\frac{(n+1)^2-1}{2}.$$

Solution 3.39. Let us prove that the converse affirmation, that is, $x^3+y^3>2$, implies that $x^2+y^3<x^3+y^4$. The power mean inequality $\sqrt{\frac{x^2+y^2}{2}}\le\sqrt[3]{\frac{x^3+y^3}{2}}$ implies that

$$x^2+y^2\le(x^3+y^3)^{2/3}\sqrt[3]{2}<(x^3+y^3)^{2/3}(x^3+y^3)^{1/3}=x^3+y^3.$$

Then $x^2-x^3<y^3-y^2\le y^4-y^3$. The last inequality follows from the fact that $y^2(y-1)^2\ge0$.

Second solution. Since $(y-1)^2\ge0$, we have that $2y\le y^2+1$, then $2y^3\le y^4+y^2$. Thus, $x^3+y^3\le x^3+y^4+y^2-y^3\le x^2+y^2$, since $x^3+y^4\le x^2+y^3$.

Solution 3.40. The inequality is equivalent to

$$(x_0-x_1)+\frac{1}{(x_0-x_1)}+(x_1-x_2)+\cdots+(x_{n-1}-x_n)+\frac{1}{(x_{n-1}-x_n)}\ge 2n.$$

Solution 3.41. Since $\frac{a+3b}{4}\ge\sqrt[4]{ab^3}$, $\frac{b+4c}{5}\ge\sqrt[5]{bc^4}$ and $\frac{c+2a}{3}\ge\sqrt[3]{ca^2}$, we can deduce that

$$(a+3b)(b+4c)(c+2a)\ge 60a^{\frac{11}{12}}b^{\frac{19}{20}}c^{\frac{17}{15}}.$$

Now prove that $c^{\frac{2}{15}}\ge a^{\frac{1}{12}}b^{\frac{1}{20}}$ or, equivalently, that $c^8\ge a^5b^3$.

Solution 3.42. We have an equivalence between the following inequalities:

$$\begin{aligned}&7(ab+bc+ca)\le 2+9\,abc\\ \Leftrightarrow\quad&7(ab+bc+ca)(a+b+c)\le 2(a+b+c)^3+9\,abc\\ \Leftrightarrow\quad&a^2b+a\,b^2+b^2c+b\,c^2+c^2a+c\,a^2\le 2(a^3+b^3+c^3.)\end{aligned}$$

For the last one use the rearrangement inequality or Tchebyshev's inequality.

Solution 3.43. Let E be the intersection of AC and BD. Then the triangles ABE and DCE are similar, which implies

$$\frac{|AB-CD|}{|AC-BD|}=\frac{|AB|}{|AE-EB|}.$$

Using the triangle inequality in ABE, we have $\frac{|AB|}{|AE-EB|}\ge1$ and we therefore conclude that $|AB-CD|\ge|AC-BD|$. Similarly, $|AD-BC|\ge|AC-BD|$.

Solution 3.44. First of all, show that $a_1 + \cdots + a_j \geq \frac{j(j+1)}{2n} a_n$, for $j \leq n$, in the following way. First, prove that the inequality is valid for $j = n$, that is, $a_1 + \cdots + a_n \geq \frac{n+1}{2} a_n$; use the fact that $2(a_1 + \cdots + a_n) = (a_1 + a_{n-1}) + \cdots + (a_{n-1} + a_1) + 2a_n$. Next, prove that if $b_j = \frac{a_1 + \cdots + a_j}{1 + \cdots + j}$, then $b_1 \geq b_2 \geq \cdots \geq b_n \geq \frac{a_n}{n}$ (to prove by induction that $b_j \geq b_{j+1}$, we need to show that $b_j \geq \frac{a_{j+1}}{j+1}$ which, on the other hand, follows from the first part for $n = j + 1$).

We provide another proof of $a_1 + \cdots + a_j \geq \frac{j(j+1)}{2n} a_n$, once again using induction. It is clear that

$$a_1 \geq a_1,$$
$$a_1 + \frac{a_2}{2} = \frac{a_1}{2} + \frac{a_1}{2} + \frac{a_2}{2} \geq \frac{a_2}{2} + \frac{a_2}{2} = a_2.$$

Now, let us suppose that the affirmation is valid for $n = 1, \ldots, j$, that is,

$$\begin{aligned} a_1 &\geq a_1 \\ a_1 + \frac{a_2}{2} &\geq a_2 \\ &\vdots \\ a_1 + \frac{a_2}{2} + \cdots + \frac{a_j}{j} &\geq a_j. \end{aligned}$$

Adding all the above inequalities, we obtain

$$ja_1 + (j-1)\frac{a_2}{2} + \cdots + \frac{a_j}{j} \geq a_1 + \cdots + a_j.$$

Adding on both sides the identity

$$a_1 + 2\frac{a_2}{2} + \cdots + j\frac{a_j}{j} = a_j + \cdots + a_1,$$

we obtain

$$(j+1)\left(a_1 + \frac{a_2}{2} + \cdots + \frac{a_j}{j}\right) \geq (a_1 + a_j) + (a_2 + a_{j-1}) + \cdots + (a_j + a_1) \geq ja_{j+1}.$$

Hence

$$a_1 + \frac{a_2}{2} + \cdots + \frac{a_j}{j} \geq \frac{j}{j+1} a_{j+1}.$$

Finally, adding $\frac{a_{j+1}}{j+1}$ on both sides of the inequality provides the final step in the induction proof.

Now,

$$\begin{aligned} a_1 + \frac{a_2}{2} + \cdots + \frac{a_n}{n} &= \frac{1}{n}(a_1 + \cdots + a_n) + \sum_{j=1}^{n-1} \left(\frac{1}{j} - \frac{1}{j+1}\right)(a_1 + \cdots + a_j) \\ &\geq \frac{1}{n}\left(\frac{n(n+1)}{2n} a_n\right) + \sum_{j=1}^{n-1} \frac{1}{j(j+1)} \frac{j(j+1)}{2n} a_n = a_n. \end{aligned}$$

Solution 3.45.

$$\begin{aligned}\left(\sum_{1\le i\le n} x_i\right)^4 &= \left(\sum_{1\le i\le n} x_i^2 + 2\sum_{1\le i<j\le n} x_ix_j\right)^2\\ &\ge 4\left(\sum_{1\le i\le n} x_i^2\right)\left(2\sum_{1\le i<j\le n} x_ix_j\right)\\ &= 8\left(\sum_{1\le i\le n} x_i^2\right)\left(\sum_{1\le i<j\le n} x_ix_j\right)\\ &= 8\sum_{1\le i<j\le n} x_ix_j(x_1^2+\cdots+x_n^2)\\ &\ge 8\sum_{1\le i<j\le n} x_ix_j(x_i^2+x_j^2).\end{aligned}$$

For the first inequality apply the AM-GM inequality.

To determine when the equality occurs, note that in the last step, two of the x_i must be different from zero and the other $n-2$ equal to zero; also in the step where the AM-GM inequality was used, the x_i which are different from zero should in fact be equal. We can prove that in such a case the constant $C = \frac{1}{8}$ is the minimum.

Solution 3.46. Setting $\sqrt[3]{a} = x$ and $\sqrt[3]{b} = y$, we need to prove that $(x^2+y^2)^3 \le 2(x^3+y^3)^2$ for $x, y > 0$.

Using the AM-GM inequality we have

$$3x^4y^2 \le x^6 + x^3y^3 + x^3y^3$$

and

$$3x^2y^4 \le y^6 + x^3y^3 + x^3y^3,$$

with equality if and only if $x^6 = x^3y^3 = y^6$ or, equivalently, if and only if $x = y$. Adding together these two inequalities and adding $x^6 + y^6$ to both sides, we get

$$x^6 + y^6 + 3x^2y^2(x^2+y^2) \le 2(x^6 + y^6 + 2x^3y^3).$$

Equality occurs when $x = y$, that is, when $a = b$.

Solution 3.47. Denote the left-hand side of the inequality as S. Since $a \ge b \ge c$ and $x \ge y \ge z$, using the rearrangement inequality we have $bz + cy \le by + cz$, then

$$(by+cz)(bz+cy) \le (by+cz)^2 \le 2((by)^2 + (cz)^2).$$

Setting $\alpha = (ax)^2$, $\beta = (by)^2$, $\gamma = (cz)^2$, we obtain

$$\frac{a^2x^2}{(by+cz)(bz+cy)} \ge \frac{a^2x^2}{2((by)^2+(cz)^2)} = \frac{\alpha}{2(\beta+\gamma)}.$$

Adding together the other two similar inequalities, we get

$$S \geq \frac{1}{2}\left(\frac{\alpha}{\beta+\gamma}+\frac{\beta}{\gamma+\alpha}+\frac{\gamma}{\alpha+\beta}\right).$$

Use Nesbitt's inequality to conclude the proof.

Solution 3.48. If XM is a median in the triangle XYZ, then $XM^2 = \frac{1}{2}XY^2 + \frac{1}{2}XZ^2 - \frac{1}{4}YZ^2$, a result of using Stewart's theorem. If we take (X, Y, Z, M) to be equal to (A, B, C, P), (B, C, D, Q), (C, D, A, R) and (D, A, B, S), and then substitute them in the formula, we then add together the four resulting equations to get a fifth equation. Multiplying both sides of the fifth equation by 4, we find that the left-hand side of the desired inequality equals $AB^2+BC^2+CD^2+DA^2+4(AC^2+BD^2)$. Thus, it is sufficient to prove that $AC^2+BD^2 \leq AB^2+BC^2+CD^2+DA^2$. This inequality is known as the "parallelogram inequality". To prove it, let O be an arbitrary point on the plane, and for each point X let x denote the vector from O to X. We expand each term in $AB^2+BC^2+CD^2+DA^2-AC^2-BD^2$, writing for instance

$$AB^2 = |a-b|^2 = |a|^2 - 2a \cdot b + |b|^2$$

and then finding that the expression equals

$$\begin{aligned}|a|^2+|b|^2+|c|^2+|d|^2 - 2(a\cdot b+b\cdot c+c\cdot d+d\cdot a-a\cdot c-b\cdot d)\\ = |a+c-b-d|^2 \geq 0,\end{aligned}$$

with equality if and only if $a+c=b+d$, that is, only if the quadrilateral $ABCD$ is a parallelogram.

Solution 3.49. Put $A = x^2+y^2+z^2$, $B = xy+yz+zx$, $C = x^2y^2+y^2z^2+z^2x^2$, $D = xyz$. Then $1 = A+2B$, $B^2 = C+2xyz(x+y+z) = C+2D$ and $x^4+y^4+z^4 = A^2-2C = 4B^2-4B+1-2C = 2C-4B+8D+1$. Then, the expression in the middle is equal to

$$3-2A+(2C-4B+8D+1) = 2+2C+8D \geq 2,$$

with equality if and only if two out of the x, y, z are zero.

Now, the right-hand expression is equal to $2+B+D$. Thus we have to prove that $2C+8D \leq B+D$ or $B-2B^2-3D \geq 0$. Using the Cauchy-Schwarz inequality, we get $A \geq B$, so that $B(1-2B) = BA \geq B^2$. Thus it is sufficient to prove that $B^2-3D = C-D \geq 0$. But $C \geq xyyz+yzzx+zxxy = D$ as can be deduced from the Cauchy-Schwarz inequality.

Solution 3.50. Suppose that $a = \frac{x}{y}$, $b = \frac{y}{z}$, $c = \frac{z}{x}$. The inequality is equivalent to

$$\left(\frac{x}{y}-1+\frac{z}{y}\right)\left(\frac{y}{z}-1+\frac{x}{z}\right)\left(\frac{z}{x}-1+\frac{y}{x}\right) \leq 1$$

and can be rewritten as $(x+z-y)(x+y-z)(y+z-x) \leq xyz$. This last inequality is valid if x, y, z are the lengths of the sides of a triangle. See Example 2.2.3.

A case remains when some out of the $u = x+z-y$, $v = x+y-z$, $w = y+z-x$ are negative. If one or three of them are negative, then the left side is negative and the inequality is clear. If two of the values u, v, w are negative, for instance u and v, then $u+v = 2x$ is also negative; but $x > 0$, so that this last situation is not possible.

Solution 3.51. First note that $abc \leq a+b+c$ implies $(abc)^2 \leq (a+b+c)^2 \leq 3(a^2+b^2+c^2)$, where the last inequality follows from inequality (1.11).

By the AM-GM inequality, $a^2+b^2+c^2 \geq 3\sqrt[3]{(abc)^2}$, then $(a^2+b^2+c^2)^3 \geq 3^3(abc)^2$. Therefore $(a^2+b^2+c^2)^4 \geq 3^2(abc)^4$.

Solution 3.52. Using the AM-GM inequality,

$$(a+b)(a+c) = a(a+b+c)+bc \geq 2\sqrt{abc(a+b+c)}.$$

Second solution. Setting $x = a+b$, $y = a+c$, $z = b+c$, and since a, b, c are positive, we can deduce that x, y, z are the side lengths of a triangle XYZ. Thus, the inequality is equivalent to $\frac{xy}{2} \geq (XYZ)$ as can be seen using the formula for the area of a triangle in Section 2.2. Now, recall that the area of a triangle with side lengths x, y, z is less than or equal to $\frac{xy}{2}$.

Solution 3.53. Since $x_i \geq 0$, then $x_i - 1 \geq -1$. Next, we can use Bernoulli's inequality for all i to get

$$(1+(x_i-1))^i \geq 1+i(x_i-1).$$

Adding these inequalities together for $1 \leq i \leq n$, gives us the result.

Solution 3.54. Subtracting 2, we find that the inequalities are equivalent to

$$0 < \frac{(a+b-c)(a-b+c)(-a+b+c)}{abc} \leq 1.$$

The left-hand side inequality is now obvious. The right-hand side inequality is Example 2.2.3.

Solution 3.55. If we prove that $\frac{a}{\sqrt{a^2+8bc}} \geq \frac{a^{4/3}}{a^{4/3}+b^{4/3}+c^{4/3}}$, it will be clear how to get the result. The last inequality is equivalent to

$$\left(a^{4/3}+b^{4/3}+c^{4/3}\right)^2 \geq a^{2/3}(a^2+8bc).$$

Apply the AM-GM inequality to each factor of

$$\left(a^{4/3}+b^{4/3}+c^{4/3}\right)^2 - \left(a^{4/3}\right)^2 = \left(b^{4/3}+c^{4/3}\right)\left(a^{4/3}+a^{4/3}+b^{4/3}+c^{4/3}\right).$$

Another method for solving this exercise is to consider the function $f(x) = \frac{1}{\sqrt{x}}$, this function is convex for $x > 0$ ($f''(x) = \frac{3}{4\sqrt{x^5}} > 0$). For $0 < a, b, c < 1$, with $a + b + c = 1$, we can deduce that $\frac{a}{\sqrt{x}} + \frac{b}{\sqrt{y}} + \frac{c}{\sqrt{z}} \geq \frac{1}{\sqrt{ax+by+cz}}$. Applying this to $x = a^2 + 8bc$, $y = b^2 + 8ca$ and $z = c^2 + 8ab$ (previously multiplying by an appropriate factor to have the condition $a + b + c = 1$), we get

$$\frac{a}{\sqrt{a^2+8bc}} + \frac{b}{\sqrt{b^2+8ca}} + \frac{c}{\sqrt{c^2+8ab}} \geq \frac{1}{\sqrt{a^3+b^3+c^3+24abc}}.$$

Also use the fact that

$$\begin{aligned}(a+b+c)^3 &= a^3+b^3+c^3+3(a^2b+a^2c+b^2a+b^2c+c^2a+c^2b)+6abc\\ &\geq a^3+b^3+c^3+24abc.\end{aligned}$$

Solution 3.56. Using the Cauchy-Schwarz inequality $\sum a_i b_i \leq \sqrt{\sum a_i^2}\sqrt{\sum b_i^2}$ with $a_i = 1$, $b_i = \frac{x_i}{1+x_1^2+x_2^2+\cdots+x_i^2}$, we can deduce that

$$\frac{x_1}{1+x_1^2} + \frac{x_2}{1+x_1^2+x_2^2} + \cdots + \frac{x_n}{1+x_1^2+\cdots+x_n^2} \leq \sqrt{n}\sqrt{\sum b_i^2}.$$

Then, it suffices to show that $\sum b_i^2 < 1$.

Observe that for $i \geq 2$,

$$\begin{aligned}b_i^2 = \left(\frac{x_i}{1+x_1^2+\cdots+x_i^2}\right)^2 &= \frac{x_i^2}{(1+x_1^2+\cdots+x_i^2)^2}\\ &\leq \frac{x_i^2}{(1+x_1^2+\cdots+x_{i-1}^2)(1+x_1^2+\cdots+x_i^2)}\\ &= \frac{1}{(1+x_1^2+\cdots+x_{i-1}^2)} - \frac{1}{(1+x_1^2+\cdots+x_i^2)}.\end{aligned}$$

For $i = 1$, use the fact that $b_1^2 \leq \frac{x_1^2}{1+x_1^2} = 1 - \frac{1}{1+x_1^2}$. Adding together these inequalities, the right-hand side telescopes to yield

$$\sum b_i^2 = \sum_{i=1}^{n}\left(\frac{x_i}{1+x_1^2+\cdots+x_i^2}\right)^2 \leq 1 - \frac{1}{1+x_1^2+\cdots+x_n^2} < 1.$$

Solution 3.57. Since there are only two possible values for α, β, γ, the three must either all be equal, or else two are equal and one is different from these two. Therefore, we have two cases to consider.

(1) $\alpha = \beta = \gamma$. In this case we have $a + b + c = 0$, and therefore

$$\begin{aligned}\left(\frac{a^3+b^3+c^3}{abc}\right)^2 &= \left(\frac{a^3+b^3-(a+b)^3}{-ab(a+b)}\right)^2\\ &= \left(\frac{(a+b)^2-a^2+ab-b^2}{ab}\right)^2 = \left(\frac{3ab}{ab}\right)^2 = 9.\end{aligned}$$

(2) Without loss of generality, we assume that $\alpha = \beta$, $\gamma \neq \alpha$, then $c = a + b$ and

$$\begin{aligned}\frac{a^3+b^3+c^3}{abc} &= \frac{a^3+b^3+(a+b)^3}{ab(a+b)} = \frac{(a+b)^2+a^2-ab+b^2}{ab}\\ &= \frac{2a^2+2b^2+ab}{ab} = 2\left(\frac{a}{b}+\frac{b}{a}\right)+1.\end{aligned}$$

If a and b have the same sign, we see that this expression is not less than 5, and its square is therefore no less than 25. If the signs of a and b are not the same, we have $\frac{a}{b}+\frac{b}{a} \leq -2$, therefore $2\left(\frac{a}{b}+\frac{b}{a}\right)+1 \leq -3$ and $\left(2\left(\frac{a}{b}+\frac{b}{a}\right)+1\right)^2 \geq 9$.

Thus, the smallest possible value is 9.

Solution 3.58. Using the AM-GM inequality, $\frac{1}{b(a+b)}+\frac{1}{c(b+c)}+\frac{1}{a(c+a)} \geq \frac{3}{XY}$, where $X = \sqrt[3]{abc}$, $Y = \sqrt[3]{(a+b)(b+c)(c+a)}$. Using AM-GM inequality again gives $X \leq \frac{a+b+c}{3}$ and $Y \leq 2\frac{a+b+c}{3}$, then $\frac{3}{XY} \geq \left(\frac{27}{2}\right)\frac{1}{(a+b+c)^2}$.

Solution 3.59. The inequality is equivalent to $a^4+b^4+c^4 \geq a^2bc+b^2ca+c^2ab$, which follows using Muirhead's theorem since $[4,0,0] \geq [2,1,1]$.

Second solution.

$$\begin{aligned}\frac{a^3}{bc}+\frac{b^3}{ca}+\frac{c^3}{ab} &= \frac{a^4}{abc}+\frac{b^4}{abc}+\frac{c^4}{abc}\\ &\geq \frac{(a^2+b^2+c^2)^2}{3abc}\\ &\geq \frac{(a+b+c)^4}{27abc} = \left(\frac{a+b+c}{3}\right)^3\frac{(a+b+c)}{abc}\\ &\geq (abc)\left(\frac{a+b+c}{abc}\right) = a+b+c.\end{aligned}$$

In the first two inequalities we applied inequality (1.11), and in the last inequality we used the AM-GM inequality.

Solution 3.60. Take $f(x)$ as $f(x) = \frac{x}{1-x}$. Since $f''(x) = \frac{2}{(1-x)^3} > 0$, $f(x)$ is convex. Using Jensen's inequality we get $f(x)+f(y)+f(z) \geq 3f(\frac{x+y+z}{3})$. But since f is increasing for $x < 1$, and because the AM-GM inequality helps us to establish that $\frac{x+y+z}{3} \geq \sqrt[3]{xyz}$, then we can deduce that $f(\frac{x+y+z}{3}) \geq f(\sqrt[3]{xyz})$.

Solution 3.61.

$$\left(\frac{a}{b+c}+\frac{1}{2}\right)\left(\frac{b}{c+a}+\frac{1}{2}\right)\left(\frac{c}{a+b}+\frac{1}{2}\right) \geq 1$$

is equivalent to $(2a+b+c)(2b+c+a)(2c+a+b) \geq 8(b+c)(c+a)(a+b)$. Now, observe that $(2a+b+c) = (a+b+a+c) \geq 2\sqrt{(a+b)(c+a)}$.

Solution 3.62. The inequality of the problem is equivalent to the following inequality:

$$\frac{(a+b-c)(a+b+c)}{c^2}+\frac{(b+c-a)(b+c+a)}{a^2}+\frac{(c+a-b)(c+a+b)}{b^2}\geq 9,$$

which in turn is equivalent to $\frac{(a+b)^2}{c^2}+\frac{(b+c)^2}{a^2}+\frac{(c+a)^2}{b^2}\geq 12$. Since $(a+b)^2\geq 4ab$, $(b+c)^2\geq 4bc$ and $(c+a)^2\geq 4ca$, we can deduce that

$$\frac{(a+b)^2}{c^2}+\frac{(b+c)^2}{a^2}+\frac{(c+a)^2}{b^2}\geq\frac{4ab}{c^2}+\frac{4bc}{a^2}+\frac{4ca}{b^2}\geq 12\sqrt[3]{\frac{(ab)(bc)(ca)}{c^2a^2b^2}}=12.$$

Solution 3.63. By the AM-GM inequality, $x^2+\sqrt{x}+\sqrt{x}\geq 3x$. Adding similar inequalities for y, z, we get $x^2+y^2+z^2+2(\sqrt{x}+\sqrt{y}+\sqrt{z})\geq 3(x+y+z)=(x+y+z)^2=x^2+y^2+z^2+2(xy+yz+zx)$.

Solution 3.64. If we multiply the equality $1=\frac{1}{a}+\frac{1}{b}+\frac{1}{c}$ by $\sqrt{abc}$, we get $\sqrt{abc}=\sqrt{\frac{ab}{c}}+\sqrt{\frac{bc}{a}}+\sqrt{\frac{ca}{b}}$. Then, it is sufficient to prove that $\sqrt{c+ab}\geq\sqrt{c}+\sqrt{\frac{ab}{c}}$. Squaring shows that this is equivalent to $c+ab\geq c+\frac{ab}{c}+2\sqrt{ab}$, $c+ab\geq c+ab(1-\frac{1}{a}-\frac{1}{b})+2\sqrt{ab}$ or $a+b\geq 2\sqrt{ab}$.

Solution 3.65. Since $(1-a)(1-b)(1-c)=1-(a+b+c)+ab+bc+ca-abc$ and since $a+b+c=2$, the inequality is equivalent to

$$0\leq(1-a)(1-b)(1-c)\leq\frac{1}{27}.$$

But $a<b+c=2-a$ implies that $a<1$ and, similarly, $b<1$ and $c<1$, therefore the left inequality is true. The other one follows from the AM-GM inequality.

Solution 3.66. It is possible to construct another triangle AA_1M with sides AA_1, A_1M, MA of lengths equal to the lengths of the medians m_a, m_b, m_c.

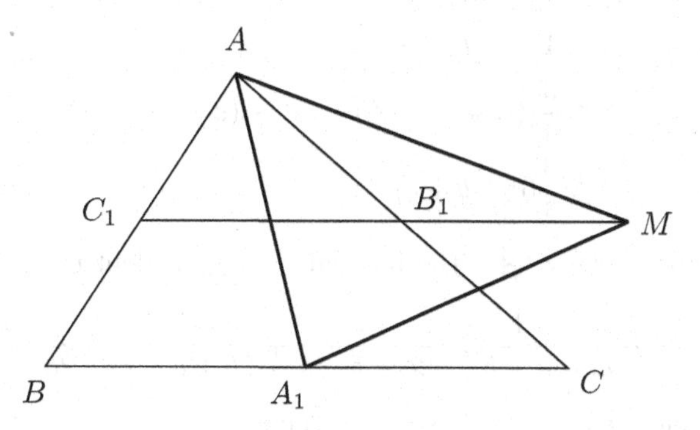

Moreover, $(AA_1M)=\frac{3}{4}(ABC)$. Then the inequality we have to prove is

$$\frac{1}{m_am_b}+\frac{1}{m_bm_c}+\frac{1}{m_cm_a}\leq\frac{3}{4}\frac{\sqrt{3}}{(AA_1M)}.$$

Now, the last inequality will be true if the triangle with side-lengths a, b, c and area S satisfies the following inequality:

$$\frac{1}{ab}+\frac{1}{bc}+\frac{1}{ca}\leq\frac{3\sqrt{3}}{4S}.$$

Or equivalently, $4\sqrt{3}\,S\leq\frac{9abc}{a+b+c}$, which is Example 2.4.6.

Solution 3.67. Substitute $cd=\frac{1}{ab}$ and $da=\frac{1}{bc}$, so that the left-hand side (LHS) inequality becomes

$$\begin{aligned}&\frac{1+ab}{1+a}+\frac{1+ab}{ab+abc}+\frac{1+bc}{1+b}+\frac{1+bc}{bc+bcd}\\&=(1+ab)\left(\frac{1}{1+a}+\frac{1}{ab+abc}\right)+(1+bc)\left(\frac{1}{1+b}+\frac{1}{bc+bcd}\right).\end{aligned}\tag{4.8}$$

Now, using the inequality $\frac{1}{x}+\frac{1}{y}\geq\frac{4}{x+y}$, we get

$$\begin{aligned}(\text{LHS})&\geq(1+ab)\frac{4}{1+a+ab+abc}+(1+bc)\frac{4}{1+b+bc+bcd}\\&=4\left(\frac{1+ab}{1+a+ab+abc}+\frac{1+bc}{1+b+bc+bcd}\right)\\&=4\left(\frac{1+ab}{1+a+ab+abc}+\frac{a+abc}{a+ab+abc+abcd}\right)=4.\end{aligned}$$

Solution 3.68. Using Stewart's theorem we can deduce that

$$l_a^2=bc\left(1-\left(\frac{a}{b+c}\right)^2\right)=\frac{bc}{(b+c)^2}((b+c)^2-a^2)\leq\frac{1}{4}((b+c)^2-a^2).$$

Using the Cauchy-Schwarz inequality leads us to

$$\begin{aligned}(l_a+l_b+l_c)^2&\leq 3(l_a^2+l_b^2+l_c^2)\\&\leq\frac{3}{4}((a+b)^2+(b+c)^2+(c+a)^2-a^2-b^2-c^2)\\&\leq\frac{3}{4}(a+b+c)^2.\end{aligned}$$

Solution 3.69. Since $\frac{1}{1-a}=\frac{1}{b+c}$, the inequality is equivalent to

$$\frac{1}{b+c}+\frac{1}{c+a}+\frac{1}{a+b}\geq\frac{2}{2a+b+c}+\frac{2}{2b+a+c}+\frac{2}{2c+a+b}.$$

Now, using the fact that $\frac{1}{x}+\frac{1}{y}\geq\frac{4}{x+y}$, we have

$$2\left(\frac{1}{b+c}+\frac{1}{c+a}+\frac{1}{a+b}\right)\geq\frac{4}{a+b+2c}+\frac{4}{b+c+2a}+\frac{4}{c+a+2b}$$

which proves the inequality.

Solution 3.70. We may take $a \leq b \leq c$. Then $c < a + b$ and

$$\frac{\sqrt[n]{2}}{2} = \frac{\sqrt[n]{2}}{2}(a+b+c) > \frac{\sqrt[n]{2}}{2}(2c) = \sqrt[n]{2c^n} \geq \sqrt[n]{b^n + c^n}.$$

Since $a \leq b$, we can deduce that

$$\begin{aligned}\left(b + \frac{a}{2}\right)^n &= b^n + nb^{n-1}\frac{a}{2} + \text{other positive terms}\\ &> b^n + \frac{n}{2}ab^{n-1} \geq b^n + a^n.\end{aligned}$$

Similarly, since $a \leq c$, we have $(c + \frac{a}{2})^n > c^n + a^n$, therefore

$$\begin{aligned}(a^n + b^n)^{\frac{1}{n}} + (b^n + c^n)^{\frac{1}{n}} + (c^n + a^n)^{\frac{1}{n}} &< b + \frac{a}{2} + \frac{\sqrt[n]{2}}{2} + c + \frac{a}{2}\\ &= a + b + c + \frac{\sqrt[n]{2}}{2} = 1 + \frac{\sqrt[n]{2}}{2}.\end{aligned}$$

Second solution. Remember that a, b, c are the lengths of the sides of a triangle if and only if there exist positive numbers x, y, z with $a = y+z$, $b = z+x$, $c = x+y$. Since $a + b + c = 1$, we can deduce that $x + y + z = \frac{1}{2}$.
Now, we use Minkowski's inequality

$$\left(\sum_{i=1}^{n}(x_i + y_i)^m\right)^{\frac{1}{m}} \leq \left(\sum_{i=1}^{n} x_i^m\right)^{\frac{1}{m}} + \left(\sum_{i=1}^{n} y_i^m\right)^{\frac{1}{m}}$$

to get

$$(a^n + b^n)^{\frac{1}{n}} = ((y+z)^n + (z+x)^n)^{\frac{1}{n}} \leq (x^n + y^n)^{\frac{1}{n}} + (2z^n)^{\frac{1}{n}} < c + \sqrt[n]{2}z.$$

Similarly, $(b^n + c^n)^{\frac{1}{n}} < a + \sqrt[n]{2}x$ and $(c^n + a^n)^{\frac{1}{n}} < b + \sqrt[n]{2}y$. Therefore

$$(a^n + b^n)^{\frac{1}{n}} + (b^n + c^n)^{\frac{1}{n}} + (c^n + a^n)^{\frac{1}{n}} < a + b + c + \sqrt[n]{2}(x + y + z) = 1 + \frac{\sqrt[n]{2}}{2}.$$

Solution 3.71. First notice that if we restrict the sums to $i < j$, then they are halved. The left-hand side sum is squared while the right-hand side sum is not, so that the desired inequality with sums restricted to $i < j$ has $(1/3)$, instead of $(2/3)$, on the right-hand side.

Consider the sum of all the $|x_i - x_j|$ with $i < j$. The number x_1 appears in $(n-1)$ terms with negative sign, x_2 appears in one term with positive sign and $(n-2)$ terms with negative sign, and so on. Thus, we get

$$-(n-1)x_1 - (n-3)x_2 - (n-5)x_3 - \cdots + (n-1)x_n = \sum(2i - 1 - n)x_i.$$

We can now apply the Cauchy-Schwarz inequality to show that the square of this sum is less than $\sum x_i^2 \sum (2i-1-n)^2$.

Looking at the sum at the other side of the desired inequality, we immediately see that it is $n\sum x_i^2 - (\sum x_i)^2$. We would like to get rid of the second term, which is easy because if we add h to every x_i the sums in the desired inequality are unaffected (since they only involve differences of the x_i), so we can choose an h to make $\sum x_i$ zero. Thus, we can finish if we can prove that $\sum(2i-1-n)^2 = \frac{n(n^2-1)}{3}$,

$$\begin{aligned}\sum(2i-1-n)^2 &= 4\sum i^2 - 4(n+1)\sum i + n(n+1)^2 \\ &= \frac{2}{3}n(n+1)(2n+1) - 2n(n+1)^2 + n(n+1)^2 \\ &= \frac{1}{3}n(n+1)(2(2n+1) - 6(n+1) + 3(n+1)) \\ &= \frac{1}{3}n(n^2-1).\end{aligned}$$

This establishes the required inequality.

Second solution. The inequality is of the Cauchy-Schwarz type, and since the problem asks us to prove that equality holds when $x_1, x_2, \dots, x_n$ form an arithmetic progression, that is, when $x_i - x_j = r(i-j)$ with $r > 0$, then consider the following inequality which is true, as can be inferred from the Cauchy-Schwarz inequality,

$$\left(\sum_{i,j} |i-j||x_i - x_j|\right)^2 \leq \sum_{i,j}(i-j)^2 \sum_{i,j}(x_i - x_j)^2.$$

Here, we already know that equality holds if and only if $(x_i - x_j) = r(i-j)$, with $r > 0$.

Since $\sum\limits_{i,j}(i-j)^2 = (2n-2)\cdot 1^2 + (2n-4)\cdot 2^2 + \dots + 2\cdot(n-1)^2 = \frac{n^2(n^2-1)}{6}$, we need to prove that $\sum\limits_{i,j}|i-j|\,|x_i - x_j| = \frac{n}{2}\sum\limits_{i,j}|x_j - x_j|$. To see that it happens compare the coefficient of x_i in each side. On the left-hand side the coefficient is

$$\begin{aligned}&(i-1) + (i-2) + \dots + (i-(i-1)) - ((i+1)-i) + ((i+2)-i) + \dots + (n-i)) \\ &\qquad = \frac{(i-1)i}{2} - \frac{(n-i)(n-i+1)}{2} = \frac{n(2i-n-1)}{2}.\end{aligned}$$

The coefficient of x_i on the right-hand side is

$$\frac{n}{2}\left(\sum_{i<j} 1 + \sum_{j>i} -1\right) = \frac{n}{2}((i-1) - (n-i)) = \frac{n(2i-n-1)}{2}.$$

Since they are equal we have finished the proof.

Solution 3.72. Let $x_{n+1} = x_1$ and $x_{n+2} = x_2$. Define

$$a_i = \frac{x_i}{x_{i+1}} \quad \text{and} \quad b_i = x_i + x_{i+1} + x_{i+2}, \quad i \in \{1, \ldots, n\}.$$

It is evident that

$$\prod_{i=1}^{n} a_i = 1, \quad \sum_{i=1}^{n} b_i = 3\sum_{i=1}^{n} x_i = 3.$$

The inequality is equivalent to

$$\sum_{i=1}^{n} \frac{a_i}{b_i} \geq \frac{n^2}{3}.$$

Using the AM-GM inequality, we can deduce that

$$\frac{1}{n}\sum_{i=1}^{n} b_i \geq \sqrt[n]{b_1 \cdots b_n} \Leftrightarrow \frac{3}{n} \geq \sqrt[n]{b_1 \cdots b_n} \Leftrightarrow \frac{1}{\sqrt[n]{b_1 \cdots b_n}} \geq \frac{n}{3}.$$

On the other hand and using again the AM-GM inequality, we get

$$\sum_{i=1}^{n} \frac{a_i}{b_i} \geq n\sqrt[n]{\frac{a_1}{b_1} \cdots \frac{a_n}{b_n}} = n\frac{\sqrt[n]{a_1 \cdots a_n}}{\sqrt[n]{b_1 \cdots b_n}} = \frac{n}{\sqrt[n]{b_1 \cdots b_n}} \geq \frac{n^2}{3}.$$

Solution 3.73. For any a positive real number, $a + \frac{1}{a} \geq 2$, with equality occurring if and only if $a = 1$. Since the numbers ab, bc and ca are non-negative, we have

$$\begin{aligned}
P(x)P\left(\frac{1}{x}\right) &= (ax^2 + bx + c)\left(a\frac{1}{x^2} + b\frac{1}{x} + c\right) \\
&= a^2 + b^2 + c^2 + ab\left(x + \frac{1}{x}\right) + bc\left(x + \frac{1}{x}\right) + ca\left(x^2 + \frac{1}{x^2}\right) \\
&\geq a^2 + b^2 + c^2 + 2ab + 2bc + 2ca = (a + b + c)^2 = P(1)^2.
\end{aligned}$$

Equality takes place if and only if either $x = 1$ or $ab = bc = ca = 0$, which in view of the condition $a > 0$ means that $b = c = 0$. Consequently, for any positive real x we have

$$P(x)P\left(\frac{1}{x}\right) \geq (P(1))^2$$

with equality if and only if either $x = 1$ or $b = c = 0$.

Second solution. Using the Cauchy-Schwarz inequality we get

$$\begin{aligned}
P(x)P\left(\frac{1}{x}\right) &= (ax^2+bx+c)\left(a\frac{1}{x^2}+b\frac{1}{x}+c\right)\\
&= \left((\sqrt{a}x)^2+(\sqrt{bx})^2+(\sqrt{c})^2\right)\left(\left(\frac{\sqrt{a}}{x}\right)^2+\left(\frac{\sqrt{b}}{\sqrt{x}}\right)^2+(\sqrt{c})^2\right)\\
&\geq \left(\sqrt{a}x\frac{\sqrt{a}}{x}+\sqrt{bx}\frac{\sqrt{b}}{\sqrt{x}}+\sqrt{c}\sqrt{c}\right)^2 = (a+b+c)^2 = (P(1))^2.
\end{aligned}$$

Solution 3.74.

$$\begin{aligned}
& \frac{a^2(b+c)+b^2(c+a)+c^2(a+b)}{(a+b)(b+c)(c+a)} \geq \frac{3}{4}\\
\Leftrightarrow\quad & \frac{a^2b+a^2c+b^2c+b^2a+c^2a+c^2b}{2abc+a^2b+a^2c+b^2c+b^2a+c^2a+c^2b} \geq \frac{3}{4}\\
\Leftrightarrow\quad & a^2b+a^2c+b^2c+b^2a+c^2a+c^2b-6abc \geq 0\\
\Leftrightarrow\quad & [2,1,0] \geq [1,1,1].
\end{aligned}$$

The last inequality follows after using Muirhead's theorem.

Second solution. Use inequality (1.11) and the Cauchy-Schwarz inequality.

Solution 3.75. Applying the AM-GM inequality to each denominator, one obtains

$$\frac{1}{1+2ab}+\frac{1}{1+2bc}+\frac{1}{1+2ca} \geq \frac{1}{1+a^2+b^2}+\frac{1}{1+b^2+c^2}+\frac{1}{1+c^2+a^2}.$$

Now, using inequality (1.11) leads us to

$$\frac{1}{1+a^2+b^2}+\frac{1}{1+b^2+c^2}+\frac{1}{1+c^2+a^2} \geq \frac{(1+1+1)^2}{3+2(a^2+b^2+c^2)} = \frac{9}{3+2\cdot 3} = 1.$$

Solution 3.76. The inequality is equivalent to each of the following ones:

$$\begin{gathered}
x^4+y^4+z^4+3(x+y+z) \geq -(x^3z+x^3y+y^3x+y^3z+z^3y+z^3x),\\
x^3(x+y+z)+y^3(x+y+z)+z^3(x+y+z)+3(x+y+z) \geq 0,\\
(x+y+z)(x^3+y^3+z^3-3xyz) \geq 0.
\end{gathered}$$

Identity (1.9) shows us that the last inequality is equivalent to

$$\frac{1}{2}(x+y+z)^2((x-y)^2+(y-z)^2+(z-x)^2) \geq 0.$$

Solution 3.77. Let O and I be the circumcenter and the incenter of the acute triangle ABC, respectively. The points O, M, X are collinear and OCX and OMC are similar right triangles. Hence we have

$$\frac{OC}{OX} = \frac{OM}{OC}.$$

Since $OC = R = OA$, we have $\frac{OA}{OM} = \frac{OX}{OA}$. Hence OAM and OXA are similar, so we have $\frac{AM}{AX} = \frac{OM}{R}$.

It now suffices to show that $OM \leq r$. Let us compare the angles $\angle OBM$ and $\angle IBM$. Since ABC is an acute triangle, O and I lie inside ABC. Now we have $\angle OBM = \frac{\pi}{2} - \angle A = \frac{1}{2}(\angle A + \angle B + \angle C) - \angle A = \frac{1}{2}(\angle B + \angle C - \angle A) \leq \frac{\angle B}{2} = \angle IBM$. Similarly, we have $\angle OCM \leq \angle ICM$. Thus the point O lies inside IBC, so we get $OM \leq r$.

Solution 3.78. Setting $a = x^2$, $b = y^2$, $c = z^2$, the inequality is equivalent to

$$x^6 + y^6 + z^6 \geq x^4yz + y^4zx + z^4xy.$$

This follows from Muirhead's theorem since $[6, 0, 0] \succeq [4, 1, 1]$.

Solution 3.79. Use the Cauchy-Schwarz inequality to see that $\sqrt{xy} + z = \sqrt{x}\sqrt{y} + \sqrt{z}\sqrt{z} \leq \sqrt{x+z}\sqrt{y+z} = \sqrt{xy + z(x+y+z)} = \sqrt{xy+z}$. Similarly, $\sqrt{yz} + x \leq \sqrt{yz+x}$ and $\sqrt{zx} + y \leq \sqrt{zx+y}$. Therefore,

$$\sqrt{xy+z} + \sqrt{yz+x} + \sqrt{zx+y} \geq \sqrt{xy} + \sqrt{yz} + \sqrt{zx} + x + y + z.$$

Solution 3.80. Using Example 1.4.11, we have

$$a^3 + b^3 + c^3 \geq \frac{(a+b+c)(a^2+b^2+c^2)}{3}.$$

Now,

$$a^3 + b^3 + c^3 \geq \frac{a+b+c}{3}(a^2+b^2+c^2) \geq \sqrt[3]{abc}(ab+bc+ca) \geq ab+bc+ca,$$

where we have used the AM-GM and the Cauchy-Schwarz inequalities.

Solution 3.81. Using Example 1.4.11, we get $(a+b+c)(a^2+b^2+c^2) \leq 3(a^3+b^3+c^3)$, but by hypothesis $a^2+b^2+c^2 \geq 3(a^3+b^3+c^3)$, hence $a+b+c \leq 1$. On the other hand,

$$4(ab+bc+ca) - 1 \geq a^2+b^2+c^2 \geq ab+bc+ca,$$

therefore $3(ab+bc+ca) \geq 1$. As

$$1 \leq 3(ab+bc+ca) \leq (a+b+c)^2 \leq 1,$$

we obtain $a+b+c = 1$. Consequently, $a+b+c = 1$ and $3(ab+bc+ca) = (a+b+c)^2$, which implies $a = b = c = \frac{1}{3}$.

Solution 3.82. The Cauchy-Schwarz inequality yields

$$(|a|+|b|+|c|)^2 \leq 3(a^2+b^2+c^2) = 9.$$

Hence $|a|+|b|+|c| \leq 3$. From the AM-GM inequality it follows that

$$a^2+b^2+c^2 \geq 3\sqrt[3]{(abc)^2}$$

or $|abc| \leq 1$, which implies $-abc \leq 1$. The requested inequality is then obtained by summation.

Solution 3.83. Notice that

$$\frac{OA_1}{AA_1} = \frac{(OBC)}{(ABC)} = \frac{OB \cdot OC \cdot BC}{4R_1} \cdot \frac{4R}{AB \cdot AC \cdot BC}.$$

Now, we have to prove that

$$OB \cdot OC \cdot BC + OA \cdot OB \cdot AB + OA \cdot OC \cdot AC \geq AB \cdot AC \cdot BC.$$

We consider the complex coordinates $O(0)$, $A(a)$, $B(b)$, $C(c)$ and obtain

$$|b| \cdot |c| \cdot |b-c| + |a| \cdot |b| \cdot |a-b| + |a| \cdot |c| \cdot |c-a| \geq |a-b| \cdot |b-c| \cdot |c-a|.$$

That is,

$$|b^2c - c^2b| + |a^2b - b^2a| + |c^2a - a^2c| \geq |ab^2 + bc^2 + ca^2 - a^2b - b^2c - c^2a|,$$

which is obvious by the triangle inequality.

Solution 3.84. Let $S = \{i_1, i_1+1, \ldots, j_1, i_2, i_2+1, \ldots, j_2, \ldots, i_p, \ldots, j_p\}$ be the ordering of S, where $j_k < i_{k+1}$ for $k = 1, 2, \ldots, p-1$. Take $S_p = a_1 + a_2 + \cdots + a_p$, $S_0 = 0$. Then

$$\sum_{i \in S} a_i = S_{j_p} - S_{i_p - 1} + S_{j_{p-1}} - S_{i_{p-1}-1} + \cdots + S_{j_1} - S_{i_1 - 1}$$

and

$$\sum_{1 \leq i \leq j \leq n} (a_i + \cdots + a_j)^2 = \sum_{0 \leq i \leq j \leq n} (S_i - S_j)^2.$$

It suffices to prove an inequality with the following form:

$$(x_1 - x_2 + \cdots + (-1)^{p+1} x_p)^2 \leq \sum_{1 \leq i < j \leq p} (x_j - x_i)^2 + \sum_{i=1}^{p} x_i^2, \tag{4.9}$$

because this means neglecting the same non-negative terms on the right-hand side of the inequality. Thus inequality (4.9) reduces to

$$4 \sum_{\substack{1 \leq i \leq j \leq p \\ j-i \text{ even}}} x_i x_j \leq (p-1) \sum_{i=1}^{p} x_i^2.$$

This can be obtained adding together inequalities with the form $4x_ix_j \le 2(x_i^2+x_j^2)$, $i<j$, $j-i$ =even (for odd i, x_i appears in such inequality $\left[\frac{p-1}{2}\right]$ times, and for even i, x_i appears in such inequality $\left[\frac{p}{2}\right]-1$ times).

Solution 3.85. Let $x=a+b+c$, $y=ab+bc+ca$, $z=abc$. Then $a^2+b^2+c^2=x^2-2y$, $a^2b^2+b^2c^2+c^2a^2=y^2-2xz$, $a^2b^2c^2=z^2$, and the inequality to be proved becomes $z^2+2(y^2-2xz)+4(x^2-2y)+8\ge 9y$ or $z^2+2y^2-4xz+4x^2-17y+8\ge 0$. Now, from $a^2+b^2+c^2\ge ab+bc+ca=y$ we obtain $x^2=a^2+b^2+c^2+2y\ge 3y$.

Also,

$$\begin{aligned} a^2b^2+b^2c^2+c^2a^2 &= (ab)^2+(bc)^2+(ca)^2 \\ &\ge ab\cdot ac+bc\cdot ab+ac\cdot bc \\ &= (a+b+c)abc = xz, \end{aligned}$$

and thus $y^2=a^2b^2+b^2c^2+c^2a^2+2xz\ge 3xz$. Hence,

$$\begin{aligned} z^2+2y^2-4xz+4x^2-17y+8 &= \left(z-\frac{x}{3}\right)^2+\frac{8}{9}(y-3)^2+\frac{10}{9}(y^2-3xz) \\ &\quad +\frac{35}{9}(x^2-3y)\ge 0, \end{aligned}$$

as required.

Second solution. Expanding the left-hand side of the inequality we obtain the equivalent inequality

$$(abc)^2+2(a^2b^2+b^2c^2+c^2a^2)+4(a^2+b^2+c^2)+8\ge 9(ab+bc+ca).$$

Since $3(a^2+b^2+c^2)\ge 3(ab+bc+ca)$ and $2(a^2b^2+b^2c^2+c^2a^2)+6\ge 4(ab+bc+ca)$ (for instance, $2a^2b^2+2\ge 4\sqrt{a^2b^2}=4ab$), it is enough to prove that

$$(abc)^2+a^2+b^2+c^2+2\ge 2(ab+bc+ca).$$

Part (i) of Exercise 1.90 tells us that it is enough to prove that $(abc)^2+2\ge 3\sqrt[3]{a^2b^2c^2}$, but this follows from the AM-GM inequality.

Solution 3.86. Let us write

$$\begin{aligned} \frac{3}{\sqrt[3]{3}}\sqrt[3]{\frac{1}{abc}+6(a+b+c)} &= \frac{3}{\sqrt[3]{3}}\sqrt[3]{\frac{1+6a^2bc+6b^2ac+6c^2ab}{abc}} \\ &= \frac{3}{\sqrt[3]{3}}\sqrt[3]{\frac{1+3ab(ac+bc)+3bc(ba+ca)+3ca(ab+bc)}{abc}}, \end{aligned}$$

and consider the condition $ab+bc+ca=1$ to obtain

$$\begin{aligned} &\frac{3}{\sqrt[3]{3}}\sqrt[3]{\frac{1+3ab-3(ab)^2+3bc-3(bc)^2+3ca-3(ca)^2}{abc}} \\ &= \frac{3}{\sqrt[3]{3}}\sqrt[3]{\frac{4-3((ab)^2+(bc)^2+(ca)^2)}{abc}}. \end{aligned}$$

It is easy to see that $3((ab)^2+(bc)^2+(ac)^2) \geq (ab+bc+ac)^2$ (use the Cauchy-Schwarz inequality). Then, it is enough to prove that

$$\frac{3}{\sqrt[3]{3}}\sqrt[3]{\frac{3}{abc}} \leq \frac{1}{abc},$$

which is equivalent to $(abc)^2 \leq \frac{1}{27}$. But this last inequality follows from the AM-GM inequality,

$$(abc)^2 = (ab)(bc)(ca) \leq \left(\frac{ab+bc+ca}{3}\right)^3 = \frac{1}{27}.$$

The equality holds if and only if $a = b = c = \frac{1}{\sqrt{3}}$.

Solution 3.87. Using symmetry, it suffices to prove that $t_1 < t_2 + t_3$. We have

$$\begin{aligned}\sum_{i=1}^n t_i \sum_{i=1}^n \frac{1}{t_i} &= n + \sum_{1\leq i<j\leq n}\left(\frac{t_i}{t_j}+\frac{t_j}{t_i}\right)\\ &= n + t_1\left(\frac{1}{t_2}+\frac{1}{t_3}\right) + \frac{1}{t_1}(t_2+t_3) + \sum_{(i,j)\neq(1,2),(1,3)}\left(\frac{t_i}{t_j}+\frac{t_j}{t_i}\right).\end{aligned}$$

Using the AM-GM inequality we get

$$\left(\frac{1}{t_2}+\frac{1}{t_3}\right) \geq \frac{2}{\sqrt{t_2t_3}},\quad t_2+t_3 \geq 2\sqrt{t_2t_3} \text{ and } \frac{t_i}{t_j}+\frac{t_j}{t_i} \geq 2 \text{ for all } i,\ j.$$

Thus, setting $a = t_1/\sqrt{t_2t_3} > 0$ and using the hypothesis, we arrive at

$$n^2+1 > \sum_{i=1}^n t_i \sum_{i=1}^n \frac{1}{t_i} \geq n + 2\frac{t_1}{\sqrt{t_2t_3}} + 2\frac{\sqrt{t_2t_3}}{t_1} + 2\left[\frac{n^2-n}{2}-2\right] = 2a + \frac{2}{a} + n^2 - 4.$$

Hence $2a+\frac{2}{a}-5<0$, which implies $1/2 < a = t_1/\sqrt{t_2t_3} < 2$. So $t_1 < 2\sqrt{t_2t_3}$, and one more application of the AM-GM inequality yields $t_1 < 2\sqrt{t_2t_3} \leq t_2+t_3$, as needed.

Solution 3.88. Note that $1+b-c = a+b+c+b-c = a+2b \geq 0$. Then

$$a\sqrt[3]{1+b-c} \leq a\left(\frac{1+1+(1+b-c)}{3}\right) = a + \frac{ab-ac}{3}.$$

Similarly,

$$\begin{aligned} b\sqrt[3]{1+c-a} &\leq b + \frac{bc-ba}{3}\\ c\sqrt[3]{1+a-b} &\leq c + \frac{ca-cb}{3}.\end{aligned}$$

Adding these three inequalities, we get

$$a\sqrt[3]{1+b-c} + b\sqrt[3]{1+c-a} + c\sqrt[3]{1+a-b} \leq a+b+c = 1.$$

Solution 3.89. If any of the numbers is zero or if an odd number of them are negative, then $x_1x_2\cdots x_6 \leq 0$ and the inequality follows.

Therefore, it can only be 2 or 4 negative numbers between the numbers in the inequality. Suppose that neither of them are zero and that there are 2 negative numbers (in the other case, change the signs of all numbers). If $y_i = |x_i|$, then it is clear that $y_1^2 + y_2^2 + \cdots + y_6^2 = 6$, $y_1 + y_2 = y_3 + \cdots + y_6$ and that $x_1x_2\cdots x_6 = y_1y_2\cdots y_6$.

From the AM-GM inequality we get

$$y_1y_2 \leq \left(\frac{y_1+y_2}{2}\right)^2 = A^2.$$

Also, the AM-GM inequality yields

$$y_3y_4y_5y_6 \leq \left(\frac{y_3+y_4+y_5+y_6}{4}\right)^4 = \left(\frac{y_1+y_2}{4}\right)^4 = \frac{1}{2^4}A^4.$$

Therefore, $y_1y_2\cdots y_6 \leq \frac{1}{2^4}A^6$.

On the other hand, the Cauchy-Schwarz inequality implies that

$$2(y_1^2+y_2^2) \geq (y_1+y_2)^2 = 4A^2$$
$$4(y_3^2+y_4^2+y_5^2+y_6^2) \geq (y_3+y_4+y_5+y_6)^2 = 4A^2.$$

Thus, $6 = y_1^2 + y_2^2 + \cdots + y_6^2 \geq 2A^2 + A^2 = 3A^2$ and then $y_1y_2\cdots y_6 \leq \frac{1}{2^4}A^6 \leq \frac{2^3}{2^4} = \frac{1}{2}$.

Solution 3.90. Use the Cauchy-Schwarz inequality with $(1,1,1)$ and $(\frac{a}{b}, \frac{b}{c}, \frac{c}{a})$ to obtain

$$(1^2+1^2+1^2)\left(\frac{a^2}{b^2}+\frac{b^2}{c^2}+\frac{c^2}{a^2}\right) \geq \left(\frac{a}{b}+\frac{b}{c}+\frac{c}{a}\right)^2.$$

The AM-GM inequality leads us to $\frac{a}{b}+\frac{b}{c}+\frac{c}{a} \geq 3\sqrt[3]{\frac{abc}{bca}} = 3$, then

$$\left(\frac{a^2}{b^2}+\frac{b^2}{c^2}+\frac{c^2}{a^2}\right) \geq \left(\frac{a}{b}+\frac{b}{c}+\frac{c}{a}\right).$$

Similarly, $\frac{a}{c}+\frac{b}{a}+\frac{c}{b} \geq 3\sqrt[3]{\frac{abc}{bca}} = 3$. Therefore,

$$\frac{a^2}{b^2}+\frac{b^2}{c^2}+\frac{c^2}{a^2}+\frac{a}{c}+\frac{b}{a}+\frac{c}{b} \geq 3+\frac{a}{b}+\frac{b}{c}+\frac{c}{a}.$$

Adding $\frac{a}{c}+\frac{b}{a}+\frac{c}{b}$ to both sides yields the result.

Solution 3.91. Note that

$$\frac{a^2+2}{2} = \frac{(a^2-a+1)+(a+1)}{2} \geq \sqrt{(a^2-a+1)(a+1)} = \sqrt{1+a^3}.$$

After substituting in the given inequality, we need to prove that

$$\frac{a^2}{(a^2+2)(b^2+2)} + \frac{b^2}{(b^2+2)(c^2+2)} + \frac{c^2}{(c^2+2)(a^2+2)} \geq \frac{1}{3}.$$

Set $x = a^2$, $y = b^2$, $z = c^2$, then $xyz = 64$ and

$$\frac{x}{(x+2)(y+2)} + \frac{y}{(y+2)(z+2)} + \frac{z}{(z+2)(x+2)} \geq \frac{1}{3}$$

if and only if

$$3[x(z+2) + y(x+2) + z(y+2)] \geq (x+2)(y+2)(z+2).$$

Now, $3(xy+yz+zx) + 6(x+y+z) \geq xyz + 2(xy+yz+zx) + 4(x+y+z) + 8$ if and only if $xy+yz+zx+2(x+y+z) \geq xyz+8 = 72$, but using the AM-GM inequality leads to $x+y+z \geq 12$ and $xy+yz+zx \geq 48$, which finishes the proof.

Solution 3.92. Observe that

$$\frac{x^5-x^2}{x^5+y^2+z^2} - \frac{x^5-x^2}{x^3(x^2+y^2+z^2)} = \frac{x^2(y^2+z^2)(x^3-1)^2}{x^3(x^5+y^2+z^2)(x^2+y^2+z^2)} \geq 0.$$

Then

$$\begin{aligned}
\sum \frac{x^5-x^2}{x^5+y^2+z^2} &\geq \sum \frac{x^5-x^2}{x^3(x^2+y^2+z^2)} \\
&= \frac{1}{x^2+y^2+z^2} \sum \left(x^2 - \frac{1}{x}\right) \\
&\geq \frac{1}{x^2+y^2+z^2} \sum (x^2 - yz) \geq 0.
\end{aligned}$$

The second inequality follows from the fact that $xyz \geq 1$, that is, $\frac{1}{x} \leq yz$. The last inequality follows from (1.8).

Second solution. First, note that

$$\frac{x^5-x^2}{x^5+y^2+z^2} = \frac{x^5+y^2+z^2-(x^2+y^2+z^2)}{x^5+y^2+z^2} = 1 - \frac{x^2+y^2+z^2}{x^5+y^2+z^2}.$$

Now we need to prove that

$$\frac{1}{x^5+y^2+z^2} + \frac{1}{x^5+z^2+x^2} + \frac{1}{x^5+x^2+y^2} \leq \frac{3}{x^2+y^2+z^2}.$$

Using the Cauchy-Schwarz inequality we get

$$(x^2+y^2+z^2)^2 \le (x^2\cdot x^3+y^2+z^2)(x^2\cdot\frac{1}{x^3}+y^2+z^2)$$

and since $xyz \ge 1$, then $x^2\cdot\frac{1}{x^3}=\frac{1}{x}\le yz$, and we have that

$$(x^2+y^2+z^2)^2 \le (x^5+y^2+z^2)(yz+y^2+z^2)$$

therefore

$$\begin{aligned}\sum\frac{1}{x^5+y^2+z^2} &\le \sum\frac{yz+y^2+z^2}{(x^2+y^2+z^2)^2} \le \sum\frac{\frac{y^2+z^2}{2}+y^2+z^2}{(x^2+y^2+z^2)^2}\\ &= \frac{3}{x^2+y^2+z^2}.\end{aligned}$$

Solution 3.93. Notice that

$$\begin{aligned}&(1+abc)\left(\frac{1}{a(b+1)}+\frac{1}{b(c+1)}+\frac{1}{c(a+1)}\right)+3\\ &=\frac{1+abc+ab+a}{a(b+1)}+\frac{1+abc+bc+b}{b(c+1)}+\frac{1+abc+ca+c}{c(a+1)}\\ &=\frac{1+a}{a(b+1)}+\frac{b(c+1)}{(b+1)}+\frac{1+b}{b(c+1)}+\frac{c(a+1)}{(c+1)}+\frac{1+c}{c(a+1)}+\frac{a(b+1)}{(a+1)}\ge 6.\end{aligned}$$

The last inequality follows after using the AM-GM inequality for six numbers.

Solution 3.94. Let R be the circumradius of the triangle ABC. Since $\angle BOC = 2\angle A$, $\angle COA = 2\angle B$ and $\angle AOB = 2\angle C$, we have that

$$\begin{aligned}(ABC)=(BOC)+(COA)+(AOB) &= \frac{R^2}{2}(\sin 2A+\sin 2B+\sin 2C)\\ &\le \frac{R^2}{2}3\sin\left(\frac{2A+2B+2C}{3}\right)\\ &= \frac{R^2}{2}3\sin\left(\frac{2\pi}{3}\right)=\frac{3\sqrt{3}R^2}{4}.\end{aligned}$$

The inequality follows since the function $\sin x$ is concave in $[0,\pi]$.

On the other hand, since BOC is isosceles, the perpendicular bisector OA' of BC is also the internal bisector of the angle $\angle BOC$, so that $\angle BOA' = \angle COA' = \angle A$; similarly $\angle COB' = \angle AOB' = \angle B$ and $\angle AOC' = \angle BOC' = \angle C$. In the triangle $B'OC'$ the altitude on the side $B'C'$ is $\frac{R}{2}$ and $B'C' = \frac{R}{2}(\tan B+\tan C)$.

Therefore, the area of the triangle $B'OC'$ is $(B'OC') = \frac{R^2}{8}(\tan B + \tan C)$. Similarly, $(C'OA') = \frac{R^2}{8}(\tan C + \tan A)$ and $(A'OB') = \frac{R^2}{8}(\tan A + \tan B)$. Then,

$$\begin{aligned}(A'B'C') = (B'OC') + (C'OA') + (A'OB') &= \frac{R^2}{4}(\tan A + \tan B + \tan C)\\ &\geq \frac{R^2}{4} 3\tan\left(\frac{A+B+C}{3}\right)\\ &= \frac{R^2}{4} 3\tan\left(\frac{\pi}{3}\right) = \frac{3\sqrt{3}R^2}{4}.\end{aligned}$$

The inequality follows since the function $\tan x$ is convex in $[0, \frac{\pi}{2}]$.

Hence,

$$(A'B'C') \geq \frac{3\sqrt{3}R^2}{4} \geq (ABC).$$

Solution 3.95. First, note that $a^2+bc \geq 2\sqrt{a^2bc} = 2\sqrt{ab}\sqrt{ca}$ and similarly $b^2+ca \geq 2\sqrt{bc}\sqrt{ab}$, $c^2+ab \geq 2\sqrt{ca}\sqrt{bc}$; then it follows that

$$\frac{1}{a^2+bc} + \frac{1}{b^2+ca} + \frac{1}{c^2+ab} \leq \frac{1}{2}\left(\frac{1}{\sqrt{ab}\sqrt{ca}} + \frac{1}{\sqrt{bc}\sqrt{ab}} + \frac{1}{\sqrt{ca}\sqrt{bc}}\right).$$

Now, using the Cauchy-Schwarz inequality in the following way

$$\left(\frac{1}{\sqrt{ab}\sqrt{ca}} + \frac{1}{\sqrt{bc}\sqrt{ab}} + \frac{1}{\sqrt{ca}\sqrt{bc}}\right)^2 \leq \left(\frac{1}{ab} + \frac{1}{bc} + \frac{1}{ca}\right)\left(\frac{1}{ca} + \frac{1}{ab} + \frac{1}{bc}\right),$$

the result follows.

Solution 3.96. From the Cauchy-Schwarz inequality we get

$$\sum_{i\neq j}\frac{a_i}{a_j}\sum_{i\neq j}a_ia_j \geq \left(\sum_{i\neq j}a_i\right)^2 = \left((n-1)\sum_{i=1}^{n}a_i\right)^2 = (n-1)^2A^2.$$

On the other hand,

$$\sum_{i\neq j}a_ia_j = \left(\sum_{i=1}^{n}a_i\right)^2 - \left(\sum_{i=1}^{n}a_i^2\right) = A^2 - A.$$

Solution 3.97. Without loss of generality, take $a_1 \leq \cdots \leq a_n$. Let $d_k = a_{k+1} - a_k$ for $k = 1, \ldots, n$. Then $d = d_1 + \cdots + d_{n-1}$. For $i < j$ we have that $|a_i - a_j| = a_j - a_i = d_i + \cdots + d_{j-1}$. Then,

$$
\begin{aligned}
s = \sum_{i<j} |a_i - a_j| &= \sum_{j=2}^{n} \sum_{i=1}^{j-1} (d_i + \cdots + d_{j-1}) \\
&= \sum_{j=2}^{n} (d_1 + 2d_2 + \cdots + (j-1)d_{j-1}) \\
&= (n-1)d_1 + (n-2)2d_2 + \cdots + 1 \cdot (n-1)d_{n-1} \\
&= \sum_{k=1}^{n-1} k(n-k)d_k.
\end{aligned}
$$

Since $k(n-k) \geq (n-1)$ (because $(k-1)(n-k-1) \geq 0$) and $4k(n-k) \leq n^2$ (from the AM-GM inequality), we obtain $(n-1)d \leq s \leq \frac{n^2 d}{4}$.

In order to see when the equality on the left holds, notice that $k(n-k) = (n-1) \Leftrightarrow n(k-1) = k^2 - 1 \Leftrightarrow k = 1$ or $k = n-1$, so that $(n-1)d = s$ only if $d_2 = \cdots = d_{n-2} = 0$, that is, $a_1 \leq a_2 = \cdots = a_{n-1} \leq a_n$.

For the second equality notice that $4k(n-k) = n^2 \Leftrightarrow k = n-k$. If n is odd, the equality $4k(n-k) = n^2$ holds only when $d_k = 0$ for all k, therefore $a_1 = \cdots = a_n = 0$. If n is even, say $n = 2k$, then only d_k can be different from zero and then $a_1 = \cdots = a_k \leq a_{k+1} = \cdots = a_{2k}$.

Solution 3.98. Consider the polynomial $P(t) = tb(t^2-b^2)+bc(b^2-c^2)+ct(c^2-t^2)$. This satisfies the identities $P(b) = P(c) = P(-b-c) = 0$, therefore $P(t) = (b-c)(t-b)(t-c)(t+b+c)$, since the coefficient of t^3 is $(b-c)$. Hence

$$
\begin{aligned}
\left|ab(a^2-b^2) + bc(b^2-c^2) + ca(c^2-a^2)\right| &= |P(a)| \\
&= |(b-c)(a-b)(a-c)(a+b+c)|.
\end{aligned}
$$

The problem is to find the least number M such that the following inequality holds for all numbers a, b, c:

$$
|(a-c)(a-b)(b-c)(a+b+c)| \leq M(a^2+b^2+c^2)^2.
$$

If (a,b,c) satisfies the inequality, then $(\lambda a, \lambda b, \lambda c)$ also satisfies it for any real number λ. Therefore, we can assume, without loss of generality, that $a^2+b^2+c^2 = 1$. In this way the problem becomes the search for the maximum value of $P = |(a-b)(a-c)(b-c)(a+b+c)|$ for real numbers a, b, c such that $a^2+b^2+c^2 = 1$.

Note that

$$
\begin{aligned}
[3(a^2+b^2+c^2)]^2 &= [2(a-b)^2 + 2(a-c)(b-c) + (a+b+c)^2]^2 \\
&\geq 8\,|(a-c)(b-c)|\,[2(a-b)^2 + (a+b+c)^2] \\
&\geq 16\sqrt{2}\,|(a-c)(b-c)(a-b)(a+b+c)| \\
&= 16\sqrt{2}P
\end{aligned}
$$

The two inequalities are obtained using the AM-GM inequality.

Thus, $P \leq \frac{9}{16\sqrt{2}}$, and the maximum value is $\frac{9}{16\sqrt{2}}$ because the equality holds with $a = \frac{3\sqrt{3}+\sqrt{6}}{6\sqrt{2}}$, $b = \frac{\sqrt{6}}{6\sqrt{2}}$ and $c = \frac{\sqrt{6}-3\sqrt{3}}{6\sqrt{2}}$.

Solution 3.99. For $a = 2$, $b = c = \frac{1}{2}$ and $n \geq 3$, the inequality is not true. If $n = 1$, the inequality becomes $abc \leq 1$, which follows from $\sqrt[3]{abc} \leq \frac{a+b+c}{3} = 1$. For the case $n = 2$, let $x = ab+bc+ca$; now since $a^2+b^2+c^2 = (a+b+c)^2 - 2(ab+bc+ca) = 9-2x$ and $x^2 = (ab+bc+ca)^2 \geq 3(a^2bc+ab^2c+abc^2) = 3abc(a+b+c) = 9abc$, the inequality is equivalent to $abc(9-2x) \leq 3$, but it will be enough to prove that $x^2(9-2x) \leq 27$. This last inequality is in turn equivalent $(2x+3)(x-3)^2 \geq 0$.

Solution 3.100. First, the AM-GM inequality leads us to $ca + c + a \geq 3\sqrt[3]{c^2a^2}$. From this we get

$$\frac{(a+1)(b+1)^2}{3\sqrt[3]{c^2a^2}+1} \geq \frac{(a+1)(b+1)^2}{ca+c+a+1} = \frac{(a+1)(b+1)^2}{(c+1)(a+1)} = \frac{(b+1)^2}{(c+1)}.$$

Similarly for the other two terms of the sum; therefore

$$\frac{(a+1)(b+1)^2}{3\sqrt[3]{c^2a^2}+1} + \frac{(b+1)(c+1)^2}{3\sqrt[3]{a^2b^2}+1} + \frac{(c+1)(a+1)^2}{3\sqrt[3]{b^2c^2}+1} \geq \frac{(b+1)^2}{(c+1)} + \frac{(c+1)^2}{(a+1)} + \frac{(a+1)^2}{(b+1)}.$$

Now, apply inequality (1.11).

Solution 3.101. Using Ravi's transformation $a = x+y$, $b = y+z$, $c = z+x$, we find that $x+y+z = \frac{3}{2}$ and $xyz \leq (\frac{x+y+z}{3})^3 = \frac{1}{8}$. Moreover,

$$\begin{aligned} a^2+b^2+c^2+\frac{4abc}{3} &= \frac{(a^2+b^2+c^2)(a+b+c)+4abc}{3} \\ &= \frac{2((y+z)^2+(z+x)^2+(x+y)^2)(x+y+z)+4(y+z)(z+x)(x+y)}{3} \\ &= \frac{4}{3}((x+y+z)^3 - xyz) \\ &\geq \frac{4}{3}\left(\left(\frac{3}{2}\right)^3 - \frac{1}{8}\right) = \frac{13}{3}. \end{aligned}$$

Therefore the minimum value is $\frac{13}{3}$.

Solution 3.102. Apply Ravi's transformation $a = y+z$, $b = z+x$, $c = x+y$, so that the inequality can be rewritten as

$$\frac{(2z)^4}{(z+x)(2x)} + \frac{(2x)^4}{(x+y)(2y)} + \frac{(2y)^4}{(y+z)(2z)} \geq (y+z)(z+x) + (z+x)(x+y) + (x+y)(y+z).$$

From inequality (1.11) and Exercise 1.27, we obtain

$$\frac{(2z)^4}{(z+x)(2x)} + \frac{(2x)^4}{(x+y)(2y)} + \frac{(2y)^4}{(y+z)(2z)} \geq \frac{8(x^2+y^2+z^2)^2}{x^2+y^2+z^2+xy+yz+zx}$$
$$\geq \frac{8(x^2+y^2+z^2)^2}{2(x^2+y^2+z^2)}.$$

On the other hand, $(y+z)(z+x)+(z+x)(x+y)+(x+y)(y+z) = 3(xy+yz+zx)+(x^2+y^2+z^2)$; then it is enough to prove that $4(x^2+y^2+z^2) \geq 3(xy+yz+zx)+(x^2+y^2+z^2)$, which can be reduced to $x^2+y^2+z^2 \geq xy+yz+zx$.

Solution 3.103. The substitution $x = \frac{a+b}{a-b}$, $y = \frac{b+c}{b-c}$, $z = \frac{c+a}{c-a}$ has the property that $xy+yz+zx = 1$. Using the Cauchy-Schwarz inequality, $(x+y+z)^2 \geq 3(xy+yz+zx) = 3$, therefore $|x+y+z| \geq \sqrt{3} > 1$.

Solution 3.104. It will be enough to consider the case $x \leq y \leq z$. Then $x = y-a$, $z = y+b$ with a, $b \geq 0$.

On the one hand, we have $xz = 1-xy-yz = 1-(y-a)y-y(y+b) = 1-2y^2+ay-by$ and on the other, $xz = (y-a)(y+b) = y^2-ay+by-ab$. Adding both identities, we get $2xz = 1-y^2-ab$, so that $2xz-1 = -y^2-ab \leq 0$. If $2xz = 1$, then $y = 0$ and $xz = 1$, a contradiction, therefore $xz < \frac{1}{2}$.

The numbers $x = y = \frac{1}{n}$ and $z = \frac{1}{2}(n-\frac{1}{n})$ satisfy $x \leq y \leq z$ and $xy+yz+zx = 1$. However, $xz = \frac{1}{2n}(n-\frac{1}{n}) = \frac{1}{2} - \frac{1}{2n^2}$ can be as close as we wish to $\frac{1}{2}$, therefore, the value $\frac{1}{2}$ cannot be improved.

Solution 3.105. Suppose that $a = [x]$ and that $r = \{x\}$. Then, the inequality is equivalent to

$$\left(\frac{a+2r}{a} - \frac{a}{a+2r}\right) + \left(\frac{2a+r}{r} - \frac{r}{2a+r}\right) > \frac{9}{2}.$$

This inequality reduces to

$$2\left(\frac{r}{a}+\frac{a}{r}\right) - \left(\frac{a}{a+2r}+\frac{r}{2a+r}\right) > \frac{5}{2}.$$

But since $\frac{r}{a}+\frac{a}{r} \geq 2$, it is enough to prove that

$$\frac{a}{a+2r}+\frac{r}{2a+r} < \frac{3}{2}.$$

But $a+2r \geq a+r$ and $2a+r \geq a+r$; moreover, the two equalities cannot hold at the same time (otherwise $a = r = 0$), therefore

$$\frac{a}{a+2r}+\frac{r}{2a+r} < \frac{a}{a+r}+\frac{r}{a+r} = 1 < \frac{3}{2}.$$

Solution 3.106. Inequality (1.11) shows that

$$a+b+c \geq \frac{1}{a}+\frac{1}{b}+\frac{1}{c} \geq \frac{3^2}{a+b+c},$$

so that $\frac{a+b+c}{3} \geq \frac{3}{a+b+c}$. Thus, it is enough to prove that $a+b+c \geq \frac{3}{abc}$.

Since $(x+y+z)^2 \geq 3(xy+yz+zx)$, we have

$$(a+b+c)^2 \geq \left(\frac{1}{a}+\frac{1}{b}+\frac{1}{c}\right)^2 \geq 3\left(\frac{1}{ab}+\frac{1}{bc}+\frac{1}{ca}\right) = \frac{3}{abc}(a+b+c),$$

and from here it is easy to conclude the proof.

Solution 3.107. By means of the Cauchy-Schwarz inequality we get

$$(a+b+1)(a+b+c^2) \geq (a+b+c)^2.$$

Then

$$\frac{a+b+c^2}{(a+b+c)^2}+\frac{a^2+b+c}{(a+b+c)^2}+\frac{a+b^2+c}{(a+b+c)^2} \geq \frac{1}{a+b+1}+\frac{1}{b+c+1}+\frac{1}{c+a+1} \geq 1.$$

Therefore,

$$2(a+b+c)+(a^2+b^2+c^2) \geq (a+b+c)^2 = a^2+b^2+c^2+2(ab+bc+ca),$$

and the result follows.

Solution 3.108. For an interior point P of ABC, consider the point Q on the perpendicular bisector of BC satisfying $AQ = AP$. Let S be the intersection of BP with the tangent to the circle at Q. Then, $SP + PC \geq SC$, therefore $BP + PC = BS + SP + PC \geq BS + SC$.

On the other hand, $BS + SC \geq BQ + QC$, then $BP + PC$ is minimum if $P = Q$.

Let T be the midpoint of MN. Since the triangle AMQ is isosceles and MT is one of its altitudes, then $MT = ZQ$ where Z is the foot of the altitude of Q over AB. Then $MN + BQ + QC = 2(MT + QC) = 2(ZQ + QC)$ is minimum when Z, Q, C are collinear and this means CZ is the altitude. By symmetry, BQ should be also an altitude and then P is the orthocenter.

Solution 3.109. Let H be the orthocenter of the triangle MNP, and let A', B', C' be the projections of H on BC, CA, AB, respectively. Since the triangle MNP is acute, H belongs to the interior of the triangle MNP; hence, it belongs to the interior of the triangle ABC too, and therefore

$$x \leq HA' + HB' + HC' \leq HM + HN + HP \leq 2X.$$

The second inequality is evident, the other two will be presented as the following two lemmas.

Lemma 1. If H is an interior point or belongs to the sides of a triangle ABC, and if A', B', C' are its projections on BC, CA, AB, respectively, then $x \leq HA' + HB' + HC'$, where x is the length of the shortest altitude of ABC.

Proof.

$$\frac{HA' + HB' + HC'}{x} \geq \frac{HA'}{h_a} + \frac{HB'}{h_b} + \frac{HC'}{h_c} = \frac{(BHC)}{(ABC)} + \frac{(CHA)}{(ABC)} + \frac{(AHB)}{(ABC)} = 1.$$

□

Lemma 2. If MNP is an acute triangle and H is its orthocenter, then $HM+HN+HP \leq 2X$, where X is the length of the largest altitude of the triangle MNP.

Proof. Suppose that $\angle M \leq \angle N \leq \angle P$, then $NP \leq PM \leq MN$ and so it happens that X is equal to the altitude MM'. We need to prove that $HM + HN + HP \leq 2MM' = 2(HM + HM')$ or, equivalently, that $HN + HP \leq HM + 2HM'$. □

Let H' be the symmetric point of H with respect to NP; since $MNH'P$ is a cyclic quadrilateral, Ptolemy's theorem tells us that

$$H'M \cdot NP = H'N \cdot MP + H'P \cdot MN \geq H'N \cdot NP + H'P \cdot NP,$$

and then we get $H'N + H'P \leq H'M = HM + 2HM'$.

Solution 3.110. Without loss of generality, we can suppose that $x \leq y \leq z$. Then $x + y \leq z + x \leq y + z$, $xy \leq zx \leq yz$, $2z^2(x + y) \geq 2y^2(z + x) \geq 2x^2(y + z)$, $\frac{1}{\sqrt{2z^2(x+y)}} \leq \frac{1}{\sqrt{2y^2(z+x)}} \leq \frac{1}{\sqrt{2x^2(y+z)}}$. If we resort to the rearrangement inequality and apply it twice, we have

$$\sum \frac{2yz}{\sqrt{2x^2(y+z)}} \geq \sum \frac{xy + zx}{\sqrt{2x^2(y+z)}}.$$

Now, adding $\sum \frac{2x^2}{\sqrt{2x^2(y+z)}}$ to both sides of the last inequality, we obtain

$$\begin{aligned}\sum \frac{2x^2 + 2yz}{\sqrt{2x^2(y+z)}} &\geq \sum \frac{2x^2 + xy + zx}{\sqrt{2x^2(y+z)}} \\ &= \sum \frac{2x^2 + x(y+z)}{\sqrt{2x^2(y+z)}} \\ &\geq \sum \frac{2\sqrt{2x^3(y+z)}}{\sqrt{2x^2(y+z)}} \\ &= 2(\sqrt{x} + \sqrt{y} + \sqrt{z}) = 2.\end{aligned}$$

Second solution. First, note that

$$\begin{aligned}\frac{x^2+yz}{\sqrt{2x^2(y+z)}} &= \frac{x^2-x(y+z)+yz}{\sqrt{2x^2(y+z)}}+\frac{x(y+z)}{\sqrt{2x^2(y+z)}}\\ &= \frac{(x-y)(x-z)}{\sqrt{2x^2(y+z)}}+\sqrt{\frac{y+z}{2}}\\ &\geq \frac{(x-y)(x-z)}{\sqrt{2x^2(y+z)}}+\frac{\sqrt{y}+\sqrt{z}}{2}.\end{aligned}$$

Similarly for the other two elements of the sum; then

$$\sum \frac{x^2+yz}{\sqrt{2x^2(y+z)}} \geq \sum \frac{(x-y)(x-z)}{\sqrt{2x^2(y+z)}}+\sqrt{x}+\sqrt{y}+\sqrt{z}.$$

Then, it is enough to prove that

$$\frac{(x-y)(x-z)}{\sqrt{2x^2(y+z)}}+\frac{(y-z)(y-x)}{\sqrt{2y^2(z+x)}}+\frac{(z-x)(z-y)}{\sqrt{2z^2(x+y)}} \geq 0.$$

Without loss of generality, suppose that $x \geq y \geq z$. Then $\frac{(x-y)(x-z)}{\sqrt{2x^2(y+z)}} \geq 0$, and

$$\begin{aligned}&\frac{(y-z)(y-x)}{\sqrt{2y^2(z+x)}}+\frac{(z-x)(z-y)}{\sqrt{2z^2(x+y)}}\\ &= \frac{(x-z)(y-z)}{\sqrt{2z^2(x+y)}}-\frac{(y-z)(x-y)}{\sqrt{2y^2(z+x)}} \geq \frac{(x-y)(y-z)}{\sqrt{2z^2(x+y)}}-\frac{(y-z)(x-y)}{\sqrt{2y^2(z+x)}}\\ &= (y-z)(x-y)\left(\frac{1}{\sqrt{2z^2(x+y)}}-\frac{1}{\sqrt{2y^2(z+x)}}\right) \geq 0.\end{aligned}$$

The last inequality is a consequence of having $y^2(z+x)=y^2z+y^2x \geq yz^2+z^2x = z^2(x+y)$.

Solution 3.111. Inequality (1.11) leads to

$$\frac{a^2}{2+b+c^2}+\frac{b^2}{2+c+a^2}+\frac{c^2}{2+a+b^2} \geq \frac{(a+b+c)^2}{6+a+b+c+a^2+b^2+c^2}.$$

Then, we need to prove that $6+a+b+c+a^2+b^2+c^2 \leq 12$, but since $a^2+b^2+c^2=3$, it is enough to prove that $a+b+c \leq 3$. But we also have $(a+b+c)^2 = a^2+b^2+c^2+2(ab+bc+ca) \leq 3(a^2+b^2+c^2) = 9$.

The equality holds if and only if $a=b=c=1$.

Solution 3.112. First, note that

$$1-\frac{a-bc}{a+bc} = \frac{2bc}{1-b-c+bc} = \frac{2bc}{(1-b)(1-c)} = \frac{2bc}{(c+a)(a+b)}.$$

Then, the inequality is equivalent to

$$\frac{2bc}{(c+a)(a+b)} + \frac{2ca}{(a+b)(b+c)} + \frac{2ab}{(b+c)(c+a)} \geq \frac{3}{2}.$$

This last inequality can be simplified to

$$4\left[bc(b+c) + ca(c+a) + ab(a+b)\right] \geq 3(a+b)(b+c)(c+a),$$

which in turn is equivalent to the inequality

$$ab + bc + ca \geq 9abc.$$

But this inequality follows from $(a+b+c)(\frac{1}{a}+\frac{1}{b}+\frac{1}{c}) \geq 9$.

Solution 3.113. Notice that $(x\sqrt{y}+y\sqrt{z}+z\sqrt{x})^2 = x^2y+y^2z+z^2x+2(xy\sqrt{yz}+yz\sqrt{zx}+zx\sqrt{xy})$.

The AM-GM inequality implies that

$$xy\sqrt{yz} = \sqrt{xyz}\sqrt{xy^2} \leq \frac{xyz+xy^2}{2},$$

then

$$(x\sqrt{y}+y\sqrt{z}+z\sqrt{x})^2 \leq x^2y+y^2z+z^2x+xy^2+yz^2+zx^2+3xyz.$$

Since $(x+y)(y+z)(z+x) = x^2y+y^2z+z^2x+xy^2+yz^2+zx^2+2xyz$, we obtain

$$\begin{aligned}(x\sqrt{y}+y\sqrt{z}+z\sqrt{x})^2 &\leq (x+y)(y+z)(z+x)+xyz\\ &\leq (x+y)(y+z)(z+x)+\frac{1}{8}(x+y)(y+z)(z+x)\\ &= \frac{9}{8}(x+y)(y+z)(z+x).\end{aligned}$$

Therefore $K^2 \geq \frac{9}{8}$, and then $K \geq \frac{3}{2\sqrt{2}}$. When $x=y=z$, the equality holds with $K = \frac{3}{2\sqrt{2}}$, hence this is the minimum value.

Second solution. Apply the Cauchy-Schwarz inequality in the following way:

$$x\sqrt{y}+y\sqrt{z}+z\sqrt{x} = \sqrt{x}\sqrt{xy}+\sqrt{y}\sqrt{yz}+\sqrt{z}\sqrt{zx} \leq \sqrt{(x+y+z)(xy+yz+zx)}.$$

After that, use the AM-GM inequality several times to produce

$$\frac{(x+y+z)}{3}\frac{(xy+yz+zx)}{3} \leq \sqrt[3]{xyz}\sqrt[3]{x^2y^2z^2} = xyz \leq \frac{(x+y)}{2}\frac{(y+z)}{2}\frac{(z+x)}{2}.$$

Solution 3.114. The left-hand side of the inequality can be written as

$$a^2b^2cd+ab^2c^2d+abc^2d^2+a^2bcd^2+a^2bc^2d+ab^2cd^2 = abcd(ab+bc+cd+ac+ad+bd).$$

The AM-GM inequality implies that $a^2b^2c^2d^2 \leq (\frac{a^2+b^2+c^2+d^2}{4})^4 = \left(\frac{1}{4}\right)^4$, hence $abcd \leq \frac{1}{16}$. To see that the factor $(ab+bc+cd+ac+ad+bd)$ is less than $\frac{3}{2}$ we can proceed in two forms.

The first way is to apply the Cauchy-Schwarz inequality to obtain

$$\begin{aligned}&(ab+bc+cd+ac+ad+bd+ba+cb+dc+ca+da+db)\\ &\quad \leq (a^2+b^2+c^2+d^2+a^2+b^2+c^2+d^2+a^2+b^2+c^2+d^2) = 3.\end{aligned}$$

The second way consists in applying the AM-GM inequality as follows:

$$\begin{aligned}&(ab+bc+cd+ac+ad+bd)\\ &\quad \leq \frac{a^2+b^2}{2}+\frac{b^2+c^2}{2}+\frac{c^2+d^2}{2}+\frac{a^2+c^2}{2}+\frac{a^2+d^2}{2}+\frac{b^2+d^2}{2} = \frac{3}{2}.\end{aligned}$$

Solution 3.115. (a) After some algebraic manipulation and some simplifications we obtain

$$\begin{aligned}&(1+x+y)^2+(1+y+z)^2+(1+z+x)^2\\ &\quad = 3+4(x+y+z)+2(xy+yz+zx)+2(x^2+y^2+z^2).\end{aligned}$$

Now, the AM-GM inequality implies that

$$\begin{aligned}(x+y+z) &\geq 3\sqrt[3]{xyz} \geq 3,\\ (xy+yz+zx) &\geq 3\sqrt[3]{x^2y^2z^2} \geq 3,\\ (x^2+y^2+z^2) &\geq 3\sqrt[3]{x^2y^2z^2} \geq 3.\end{aligned}$$

Then, $(1+x+y)^2+(1+y+z)^2+(1+z+x)^2 \geq 3+4\cdot 3+2\cdot 3+2\cdot 3 = 27$.

The equality holds when $x=y=z=1$.

(b) Again, after simplification, the inequality is equivalent to

$$\begin{aligned}&3+4(x+y+z)+2(xy+yz+zx)+2(x^2+y^2+z^2)\\ &\quad \leq 3(x^2+y^2+z^2)+6(xy+yz+zx)\end{aligned}$$

and also to $3+4u \leq u^2+2v$, where $u=x+y+z\geq 3$ and $v=xy+yz+zx\geq 3$. But $u\geq 3$ implies that $(u-2)^2\geq 1$, then $(u-2)^2+2v\geq 1+6=7$.

The equality holds when $u=3$ and $v=3$, that is, when $x=y=z=1$.

Solution 3.116. Notice that

$$\frac{1}{1+a^2(b+c)} = \frac{1}{1+a(ab+ac)} = \frac{1}{1+a(3-bc)} = \frac{1}{3a+1-abc}.$$

The AM-GM inequality implies that $1 = \frac{ab+bc+ca}{3} \geq \sqrt[3]{a^2b^2c^2}$, then $abc \leq 1$. Thus

$$\frac{1}{1+a^2(b+c)} = \frac{1}{3a+1-abc} \leq \frac{1}{3a}.$$

Similarly, $\frac{1}{1+b^2(c+a)} \leq \frac{1}{3b}$ and $\frac{1}{1+c^2(a+b)} \leq \frac{1}{3c}$. Therefore,

$$\begin{aligned}\frac{1}{1+a^2(b+c)} + \frac{1}{1+b^2(c+a)} + \frac{1}{1+c^2(a+b)} &\leq \frac{1}{3a} + \frac{1}{3b} + \frac{1}{3c} \\ &= \frac{bc+ca+ab}{3abc} = \frac{1}{abc}.\end{aligned}$$

Solution 3.117. The inequality is equivalent to

$$(a+b+c)\left(\frac{1}{a+b} + \frac{1}{b+c} + \frac{1}{c+a}\right) \geq k + (a+b+c)k = (a+b+c+1)k.$$

On the other hand, using the condition $a+b+c = ab+bc+ca$, we have

$$\begin{aligned}\frac{1}{a+b} + \frac{1}{b+c} + \frac{1}{c+a} &= \frac{a^2+b^2+c^2+3(ab+bc+ca)}{(a+b)(b+c)(c+a)} \\ &= \frac{a^2+b^2+c^2+2(ab+bc+ca)+(ab+bc+ca)}{(a+b)(b+c)(c+a)} \\ &= \frac{(a+b+c)(a+b+c+1)}{(a+b+c)^2-abc}.\end{aligned}$$

Hence

$$\frac{(a+b+c)}{(a+b+c+1)}\left(\frac{1}{a+b} + \frac{1}{b+c} + \frac{1}{c+a}\right) = \frac{(a+b+c)^2}{(a+b+c)^2-abc} \geq 1,$$

and since the equality holds if and only if $abc = 0$, we can conclude that $k = 1$ is the maximum value.

Solution 3.118. Multiplying both sides of the inequality by the factor $(a+b+c)$, we get the equivalent inequality

$$9(a+b+c)(a^2+b^2+c^2) + 27abc \geq 4(a+b+c)^3,$$

which in turn is equivalent to the inequality

$$5(a^3+b^3+c^3) + 3abc \geq 3(ab(a+b) + ac(a+c) + bc(b+c)).$$

By the Schür inequality with $n = 1$, Exercise 1.83, it follows that

$$a^3+b^3+c^3+3abc \geq ab(a+b) + bc(b+c) + ca(c+a),$$

and the Muirhead's inequality tells us that $2[3,0,0] \geq 2[2,1,0]$, which is equivalent to

$$4(a^3+b^3+c^3) \geq 2(ab(a+b) + ac(a+c) + bc(b+c)).$$

Adding these last inequalities, we get the result.

Solution 3.119. Lemma. If $a, b > 0$, then $\frac{1}{(a-b)^2} + \frac{1}{a^2} + \frac{1}{b^2} \geq \frac{4}{ab}$.

Proof. In order to prove the lemma notice that $\frac{1}{(a-b)^2} + \frac{1}{a^2} + \frac{1}{b^2} - \frac{4}{ab} = \frac{(a^2+b^2-3ab)^2}{a^2b^2(a-b)^2}$. □

Without loss of generality, $z = \min\{x, y, z\}$; now apply the lemma with $a = (x - z)$ and $b = (y - z)$, to obtain

$$\frac{1}{(x-y)^2} + \frac{1}{(y-z)^2} + \frac{1}{(z-x)^2} \geq \frac{4}{(x-z)(y-z)}.$$

Now, it is left to prove that $xy + yz + zx \geq (x - z)(y - z)$; but this is equivalent to $2z(y + x) \geq z^2$, which is evident.

Solution 3.120. In the case of part (i), there are several ways to prove it.

First form. We can prove that

$$\frac{x^2}{(x-1)^2} + \frac{y^2}{(y-1)^2} + \frac{z^2}{(z-1)^2} - 1 = \frac{(yz + zx + xy - 3)^2}{(x-1)^2(y-1)^2(z-1)^2}.$$

Second form. With the substitution $a = \frac{x}{x-1}$, $b = \frac{y}{y-1}$, $c = \frac{z}{z-1}$, the inequality is equivalent to $a^2 + b^2 + c^2 \geq 1$, and the condition $xyz = 1$ is equivalent to $abc = (a - 1)(b - 1)(c - 1)$ or $(ab + bc + ca) + 1 = a + b + c$. With the previous identities we can obtain

$$\begin{aligned} a^2 + b^2 + c^2 &= (a + b + c)^2 - 2(ab + bc + ca) \\ &= (a + b + c)^2 - 2(a + b + c - 1) \\ &= (a + b + c - 1)^2 + 1, \end{aligned}$$

therefore

$$a^2 + b^2 + c^2 = (a + b + c - 1)^2 + 1.$$

Part (ii) can be proved depending on how we prove part (i). For instance, if we used the *second form*, the equality holds when $a^2 + b^2 + c^2 = 1$ and $a + b + c = 1$. (In the *first form*, the equality holds when $xyz = 1$ and $xy + yz + zx = 3$). From the equations we can cancel out one variable, for instance c (and since $c = 1 - a - b$, if we find that a and b are rational numbers, then c will be a rational number too), to obtain $a^2 + b^2 + ab - a - b = 0$, an identity that we can think of as a quadratic equation in the variable b with roots $b = \frac{1-a\pm\sqrt{(1-a)(1+3a)}}{2}$, which will be rational numbers if $(1 - a)$ and $(1 + 3a)$ are squares of rational numbers. If $a = \frac{k}{m}$, then $m - k$ and $m + 3k$ are squares of integers, for instance, if $m = (k - 1)^2 + k$, then $m - k = (k - 1)^2$ and $m + 3k = (k + 1)^2$. Thus, the rational numbers $a = \frac{k}{m}$, $b = \frac{m-k+k^2-1}{2m}$ and $c = 1 - a - b$, when k varies in the integer numbers, are rational numbers where the equality holds. There are some exceptions, that is, when $k = 0$, 1, since the values $a = 0$ or 1 are not allowed.

Notation

We use the following standard notation:

$\mathbb{N}$	the positive integers (natural numbers)
$\mathbb{R}$	the real numbers
$\mathbb{R}^+$	the positive real numbers
$\Leftrightarrow$	iff, if and only if
$\Rightarrow$	implies
$a \in A$	the element a belongs to the set A
$A \subset B$	A is a subset of B
$\|x\|$	the absolute value of the real number x
$\{x\}$	the fractional part of the real number x
$[x]$	the integer part of the real number x
$[a, b]$	the set of real numbers x such that $a \leq x \leq b$
(a, b)	the set of real numbers x such that $a < x < b$
$f : [a, b] \to \mathbb{R}$	the function f defined in $[a, b]$ with values in $\mathbb{R}$
$f'(x)$	the derivative of the function $f(x)$
$f''(x)$	the second derivative of the function $f(x)$
$\det A$	the determinant of the matrix A
$\sum_{i=1}^{n} a_i$	the sum $a_1 + a_2 + \cdots + a_n$
$\prod_{i=1}^{n} a_i$	the product $a_1 \cdot a_2 \cdots a_n$
$\prod_{i \neq j} a_i$	the product of all $a_1, a_2, \ldots, a_n$ except a_j
$\max\{a, b, \ldots\}$	the maximum value between $a, b, \ldots$
$\min\{a, b, \ldots\}$	the minimum value between $a, b, \ldots$
$\sqrt{x}$	the square root of the positive real number x
$\sqrt[n]{x}$	the n-th root of the real number x
$\exp x = e^x$	the exponential function
$\sum_{\text{cyclic}} f(a, b, \ldots)$	represents the sum of the function f evaluated in all cyclic permutations of the variables $a, b, \ldots$

We use the following notation for the section of Muirhead's theorem:

$\sum_! F(x_1,\ldots,x_n)$	the sum of the $n!$ terms obtained from evaluating F in all possible permutations of $(x_1,\ldots,x_n)$
$(b) \prec (a)$	(b) is majorized by (a)
$[b] \leq [a]$	$\frac{1}{n!}\sum_! x_1^{b_1}x_2^{b_2}\cdots x_n^{b_n} \leq \frac{1}{n!}\sum_! x_1^{a_1}x_2^{a_2}\cdots x_n^{a_n}$.

We use the following geometric notation:

A, B, C	the vertices of the triangle ABC
a, b, c	the lengths of the sides of the triangle ABC
A', B', C'	the midpoints of the sides BC, CA and AB
$\angle ABC$	the angle ABC
$\angle A$	the angle in the vertex A or the measure of the angle A
(ABC)	the area of the triangle ABC
$(ABCD...)$	the area of the polygon $ABCD...$
m_a, m_b, m_c	the lengths of the medians of the triangle ABC
h_a, h_b, h_c	the lengths of the altitudes of the triangle ABC
l_a, l_b, l_c	the lengths of the internal bisectors of the triangle ABC
s	the semiperimeter of the triangle ABC
r	the inradius of the triangle ABC, the radius of the incircle
R	the circumradius of the triangle ABC, the radius of the circumcircle
I, O, H, G	the incenter, circumcenter, orthocenter and centroid of the triangle ABC
I_a, I_b, I_c	the centers of the excircles of the triangle ABC.

We use the following notation for reference of problems:

IMO	International Mathematical Olympiad
APMO	Asian Pacific Mathematical Olympiad
(country, year)	problem corresponding to the mathematical olympiad celebrated in that country, in that year, in some stage.

Bibliography

[1] Altshiller, N., *College Geometry: An Introduction to Modern Geometry of the Triangle and the Circle.* Barnes and Noble, 1962.

[2] Andreescu, T., Feng, Z., *Problems and Solutions from Around the World.* Mathematical Olympiads 1999-2000. MAA, 2002.

[3] Andreescu, T., Feng, Z., Lee, G., *Problems and Solutions from Around the World.* Mathematical Olympiads 2000-2001. MAA, 2003.

[4] Andreescu, T., Enescu, B., *Mathematical Olympiad Treasures.* Birkhäuser, 2004.

[5] Barbeau, E.J., Shawyer, B.L.R., *Inequalities.* A Taste of Mathematics, vol. 4, 2000.

[6] Bulajich, R., Gómez Ortega, J.A., *Geometría.* Cuadernos de Olimpiadas de Matemáticas, Instituto de Matemáticas, UNAM, 2002.

[7] Bulajich, R., Gómez Ortega, J.A., *Geometría. Ejercicios y Problemas.* Cuadernos de Olimpiadas de Matemáticas, Instituto de Matemáticas, UNAM, 2002.

[8] Courant, R., Robbins, H., *¿Qué son las Matemáticas?* Fondo de Cultura Económica, 2002.

[9] Coxeter, H., Greitzer, S., *Geometry Revisited.* New Math. Library, MAA, 1967.

[10] Dorrie, H., *100 Great Problems of Elementary Mathematics.* Dover, 1965.

[11] Engel, A.,*Problem-Solving Strategies.* Springer-Verlag, 1998.

[12] Fomin, D., Genkin, S., Itenberg, I., *Mathematical Circles.* Mathematical World, Vol. 7. American Mathematical Society, 1996.

[13] Hardy, G.H., Littlewood, J.E., Pòlya, G., *Inequalities.* Cambridge at the University Press, 1967.

[14] Honsberger, R., *Episodes in Nineteenth and Twentieth Century Euclidean Geometry.* New Math. Library, MAA, 1995.

[15] Kazarinoff, N., *Geometric Inequalities.* New Math. Library, MAA. 1961.

[16] Larson, L., *Problem-Solving Through Problems.* Springer-Verlag, 1990.

[17] Mitrinovic, D., *Elementary Inequalities.* Noordhoff Ltd., Groningen, 1964.

[18] Niven, I., *Maxima and Minima Without Calculus.* The Dolciani Math. Expositions, MAA, 1981.

[19] Shariguin, I., *Problemas de Geometría.* Editorial Mir, 1989.

[20] Soulami, T., *Les Olympiades de Mathématiques.* Ellipses, 1999.

[21] Spivak, M., *Calculus.* Editorial Benjamin, 1967.

Index

Absolute value, 2

Concavity
    Geometric interpretation, 25
Convexity
    Geometric interpretation, 25

discrepancy, 46

Erdős-Mordell
    theorem, 81–84, 88
Euler
    theorem, 66

Fermat
    point, 90, 92
Function
    concave, 23
    convex, 20
    quadratic, 4

Greater than, 1

Inequality
    arithmetic mean–geometric mean, 9, 47
        weighted, 27
    Bernoulli, 31
    Cauchy-Schwarz, 15, 35
        Engel form, 35
    Euler, 67
    Hölder, 27
        generalized, 32
    harmonic mean–geometric mean, 8
    helpful, 34
    Jensen, 21
    Leibniz, 69
    Minkowski, 28
    Nesbitt, 16, 37, 65
    Popoviciu, 32
    power mean, 32
    Ptolemy, 53
    quadratic mean–arithmetic mean, 19, 36
    rearrangement, 13
    Schur, 31
    Tchebyshev, 18
    triangle, 3
        general form, 3
    Young, 27

Leibniz
    theorem, 68

Mean
    arithmetic, 7, 9, 19, 31
    geometric, 7, 9, 19, 31
    harmonic, 8, 19
    power, 32
    quadratic, 19
Muirhead
    theorem, 43, 44

Ortic
    triangle, 95, 98

Pappus
    theorem, 80
Pedal
    triangle, 99

Problem
  Fagnano, 88, 94
  Fermat-Steiner, 88
  Heron, 92
    with a circle, 93
  Pompeiu, 53

Real line, 1

Smaller than, 1
Smaller than or equal to, 2
Solution
  Fagnano problem
    Fejér L., 96
    Schwarz H., 96
  Fermat-Steiner problem
    Hofmann-Gallai, 91
    Steiner, 92, 94
    Torricelli, 88, 90

Transformation
  Ravi, 55, 73

Viviani
  lemma, 88, 90